Wind Power - From Energy Conversion to Technological and Operational Challenges

Edited by Basel I. Abed Ismail

Published in London, United Kingdom

Wind Power - From Energy Conversion to Technological and Operational Challenges
http://dx.doi.org/10.5772/intechopen.1008573
Edited by Basel I. Abed Ismail

Contributors
Basel I. Abed Ismail, Cedric Caruana, Claire Grima, Cristian Helera, Dan Alexandru Stoichescu, Fred Nitzsche, Monica Borunda

First published in London, United Kingdom, 2026 by IntechOpen
IntechOpen is the global imprint of INTECHOPEN LIMITED, registered in England and Wales, registration number: 11086078, 167-169 Great Portland Street, London, W1W 5PF, United Kingdom

For EU product safety concerns: IN TECH d.o.o., Prolaz Marije Krucifikse Kozulić 3, 51000 Rijeka, Croatia, info@intechopen.com or visit our website at intechopen.com.

British Library Cataloguing-in-Publication Data
A catalogue record for this book is available from the British Library

Wind Power - From Energy Conversion to Technological and Operational Challenges
Edited by Basel I. Abed Ismail
p. cm.
Print ISBN 978-1-83635-593-9
Online ISBN 978-1-83635-592-2
eBook (PDF) ISBN 978-1-83635-594-6

Meet the editor

Dr. Basel Ismail is a successful long-track experienced academician, Professor and former (2016-2024) Chair of the Department of Mechanical & Mechatronics Engineering at Lakehead University, pioneering and international academic researcher in alternative energy and renewable & green energy engineering technologies, an academic editor for many books, an author for many peer reviewed articles and book chapters, a member of the Canadian Society of Mechanical Engineering (CSME), and a licensed Professional Engineer (P. Eng.) of Ontario, Canada.

Contents

Preface

Wind power generation is the power produced from the local and planetary movement of the wind. It is a renewable and green energy source. Wind power generation is one of the fastest-evolving sources of renewable energy and has been making a significant contribution to large-scale electrical power production in many countries across the globe. At the end of 2023, it was estimated that wind power generation capacity exceeded 1 TW (1000 GW). This book, *Wind Power – From Energy Conversion to Technological and Operational Challenges*, offers and discusses unique topics in modern wind power technologies. The book presents diverse chapters covering fundamental wind energy conversion aspects, with illustrative numerical case studies, sophisticated control strategies, and the complex integration challenges of wind power systems, as well as innovative methods for enhancing the utilization of wind resources using Artificial Intelligence (AI) techniques.

Furthermore, it delves into emerging concepts such as vertical-axis wind turbines (VAWTs), intelligent farms, and other interesting topics. This book is designed to be an essential source of information and a foundation for extended research for academicians, researchers, policymakers, and industry engineering practitioners dedicated to advancing modern wind power technologies. The Editor extends sincere gratitude to all contributing authors for their expertise and to Ms. Kristina Kardum Cvitan at IntechOpen for her exceptional management of the publication process.

Dr. Basel I. Abed Ismail, P.Eng.
Professor and Former Chair (2016-2024),
Department of Mechanical and Mechatronics Engineering,
Lakehead University,
Thunder Bay, Ontario, Canada

Chapter 1

Perspective Chapter: Fundamental Wind Energy Conversion Aspects with Illustrative Numerical Examples

Basel I. Abed Ismail

Abstract

Wind energy is the process of converting the kinetic energy of the wind into electrical power using wind turbines, typically arranged into wind farms for large power generation. It is naturally replenishing and green energy resource, thus being a renewable energy resource, which has shown significant technological progress and spread worldwide as a successful source of renewable energy for producing electricity. It is becoming one of the fastest growing renewable energy sources and is a key component in the global transition to sustainable power generation. It has also been globally recognized as a critical technology for conserving natural resources, mitigating greenhouse gas emissions and air pollution, and fulfilling the rising global demand for primary electrical energy across diverse sectors. In this chapter, wind turbines were briefly introduced and described, with the main goal to present and discuss fundamental aspects of 1-D energy formulation using wind energy principles. It was found through detailed 1-D energy formulations that the maximum wind turbine conversion efficiency is approximately 59.3%, which confirms the Betz limit. Furthermore, this chapter discusses illustrative numerical examples that are grounded in the fundamentals of energy conversion and modeling, with practical applications provided.

Keywords: power generation, wind energy, Betz limit, wind turbine efficiency, wind available energy, wind attainable power, energy conversion, WECS

1. Introduction

1.1 Wind characteristics

Wind energy is fundamentally an indirect form of solar energy, because the wind itself is primarily caused by the Sun's uneven heating of the Earth's surface. The wind phenomena can be categorized into two main types: planetary and local winds [1].

Planetary winds:

Planetary winds are large-scale, global air movements driven by the significant difference in solar heating between the Equator and the Poles.

- The Sun heats the surface much more intensely near the Equator, causing the warm, tropical air to rise and flow toward the Poles in the upper atmosphere.
- Concurrently, colder, denser air from the Poles flows back toward the Equator closer to the Earth's surface.
- The Earth's rotation influences the direction of this planetary air movement. The wind's velocity and direction also exhibit seasonal variations based on the amount of solar energy received.

Local winds:

Local winds are smaller-scale air movements driven by two principal mechanisms:

A. Variation of heating between water and land:

- During the day, solar energy rapidly converts into heat on the land surface.
- In contrast, water absorbs solar energy more slowly and uses some of it for evaporation, causing the land to become significantly hotter than adjacent water bodies.
- A pressure gradient is established as warm air rises over the land, inducing the advection of cooler, denser marine air inland, a phenomenon known as a sea or lake breeze.

B. Heating and cooling of mountain slopes:

- Air above hill and mountain slopes heats up and cools down more rapidly than the air over the lower-lying valleys.
- During the day, the heated air on the slopes rises (anabatic wind).
- At night, the cool, dense air flows down the slopes into the valleys (katabatic wind).

1.2 Classification of wind turbines

Wind turbines are wind energy electromechanical conversion devices that utilize available wind energy to produce electrical power. Wind turbines are broadly classified as (1) horizontal-axis wind turbines (HAWT), shown in **Figure 1**, and (2) vertical-axis wind turbines (VAWT), shown in **Figure 2** [2].

In terms of their rated power capacity, wind turbines are classified as small or large. Small wind turbines are those of less than 100 kW rated capacity. They are generally used for generating power form local applications in a particular location. Large wind turbines are those of 100 kW capacity or greater. They are used to generate power for distribution in central power grids. **Figure 3** shows modern

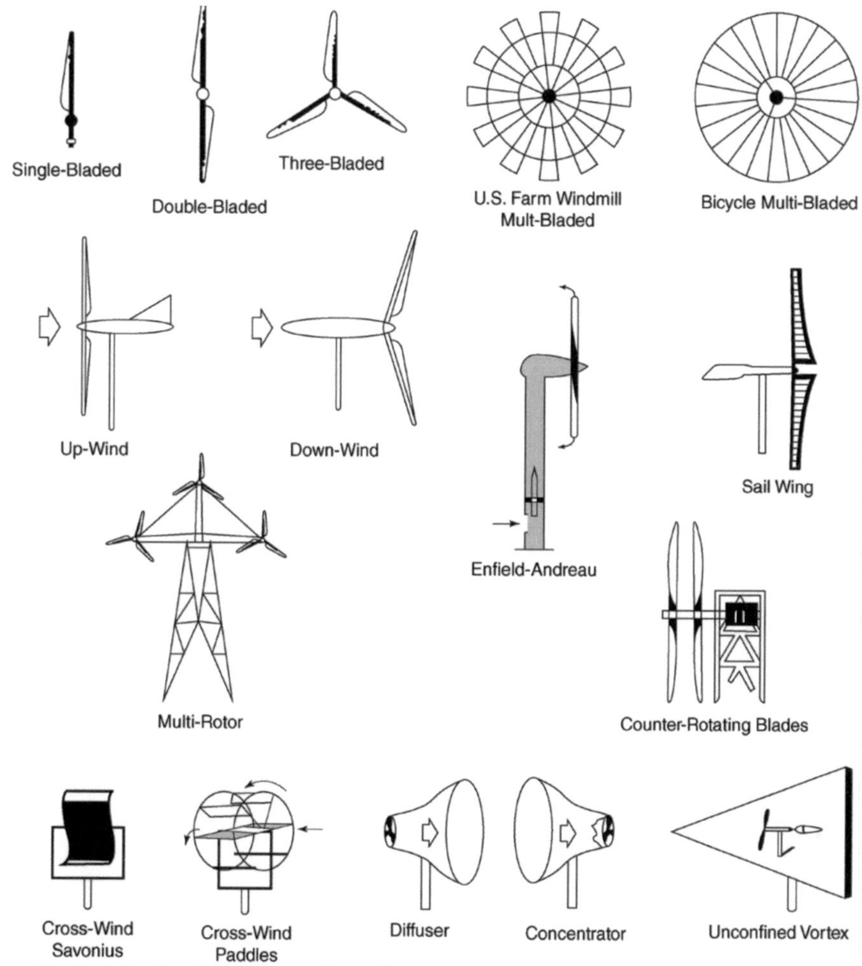

Figure 1.
Types of horizontal-axis wind turbines (HAWT) [2].

designs of HAWT and VAWT systems [3] and **Figure 4** [4] and **Figure 5** [3] show the main components of wind energy conversion system (WECS) and HAWT [3, 4]. They consist of several main components that work together to achieve this conversion.

The key components can be grouped into systems for capture, conversion, support, and control.

1. Rotor and blades (capture system)

The Rotor is the rotating part of the turbine, consisting of the Hub and the Blades.

- Blades: These aerodynamic structures capture the wind's kinetic energy. They are designed like an airplane wing, using lift to cause the rotor to spin.
- Hub: This central piece connects the blades to the low-speed shaft.

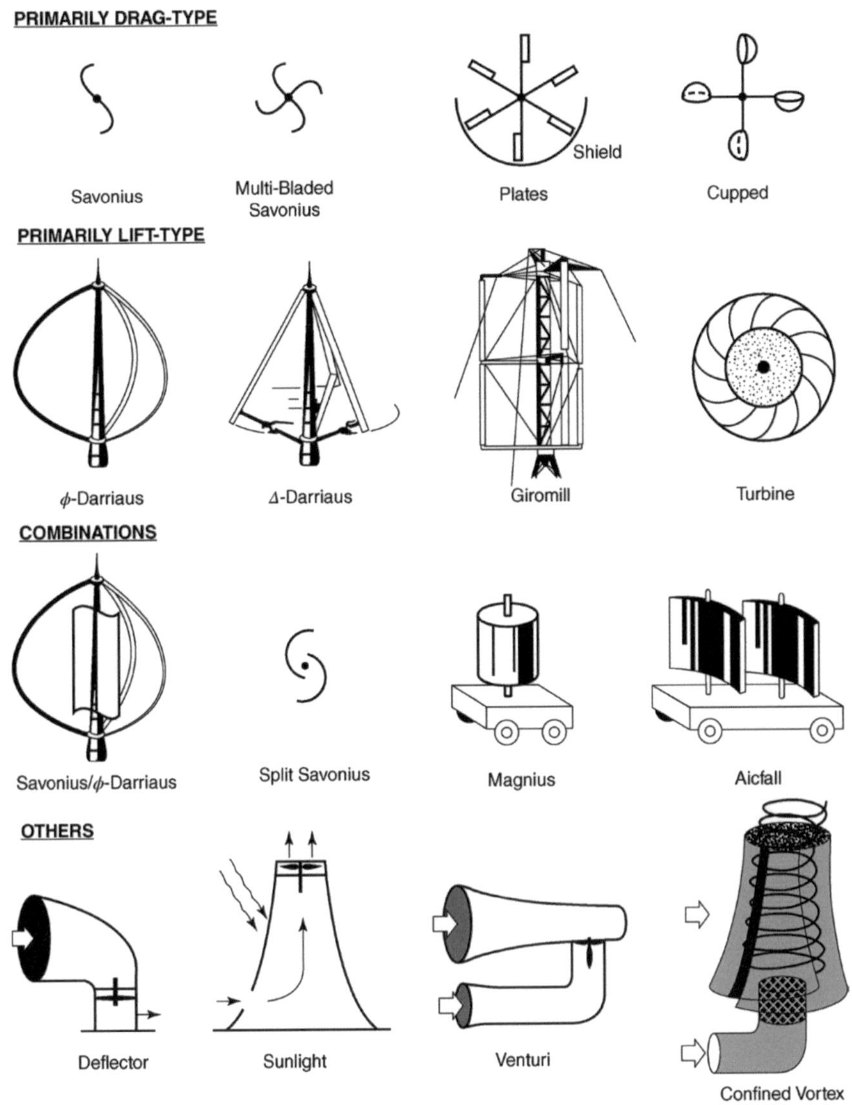

Figure 2.
Types of vertical-axis wind turbines (VAWT) [2].

2. Nacelle and drivetrain (conversion system)

The Nacelle is the housing located at the top of the tower, which protects the major power generation components. This is where the mechanical energy is converted to electrical energy.

- Low-speed shaft: Connects the rotor hub to the gearbox and spins at the same rate as the blades (typically 8–20 revolutions per minute, or RPM).

- Gearbox: (Not used in all turbines, like direct-drive models). The gearbox takes the slow, high-torque rotation of the low-speed shaft and “steps up” the speed to the high RPM needed by the generator (e.g., 1000–1800 RPM).

DOI: http://dx.doi.org/10.5772/intechopen.1014078

Figure 3.
Modern designs of HAWT and VAWT systems [3].

- High-speed shaft: Connects the gearbox to the generator.
- Generator: This is the core electrical workhorse. It uses the rotational mechanical energy from the shaft to produce electrical energy (electricity) through electromagnetic induction.

3. Structural components (support system)

These components provide the necessary height and stability for the system.

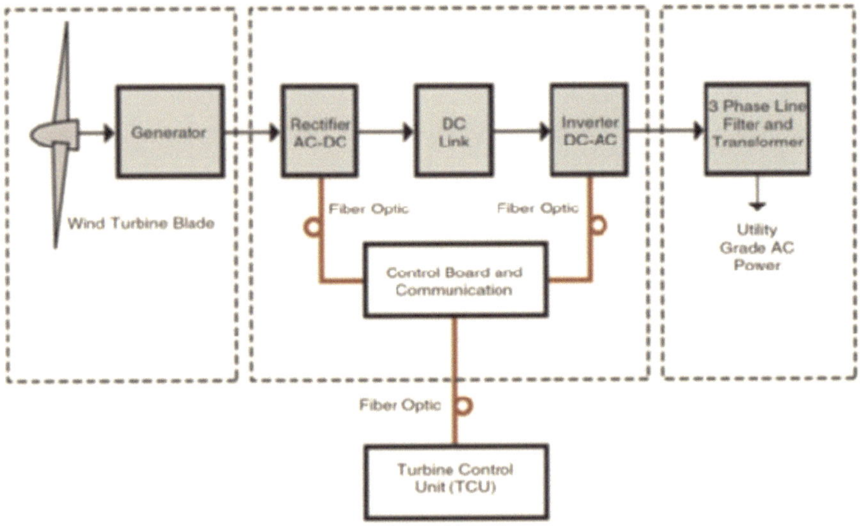

Figure 4.
Main components of a WECS [4].

- Tower: The structure (often tubular steel) that supports the nacelle and rotor. Its height is crucial, as wind speeds are generally stronger and more consistent farther from the ground.

- Foundation: The base, usually made of reinforced concrete, that anchors the tower to the ground, ensuring structural integrity and stability against high winds.

4. Control and orientation systems

These systems optimize power generation and ensure the turbine operates safely.

- Yaw system (wind orientation control): Motors and gears that rotate the entire nacelle on top of the tower so that the rotor assembly is always facing directly into the wind, maximizing energy capture.

- Pitch system: Adjusts the angle (pitch) of the blades to control the rotor speed and optimize power output. In very high winds, it can "feather" the blades (turn them out of the wind) to stop the rotor and prevent damage.

- Controller: A computer system that continuously monitors the turbine's conditions (e.g., wind speed, temperature) and sends commands to the yaw and pitch systems.

- Brake: A mechanical or electrical brake system to stop the rotor in an emergency or during maintenance.

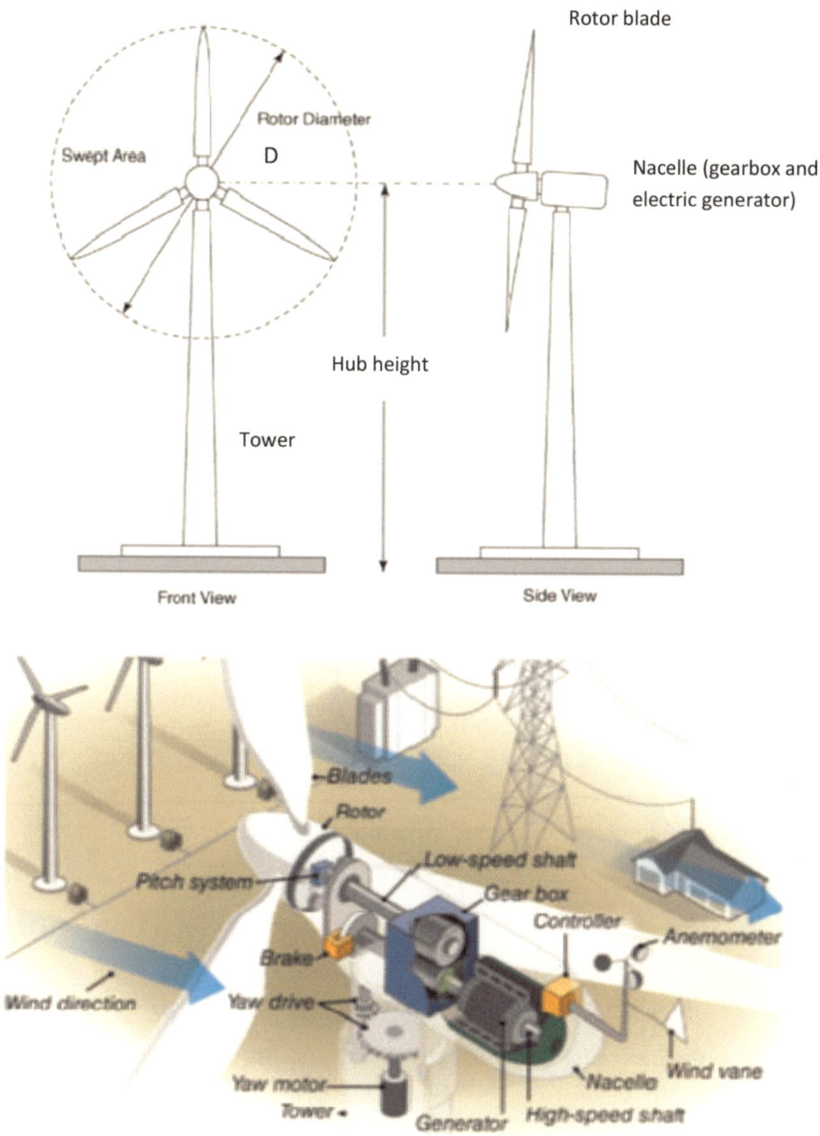

Figure 5.
Schematic diagram showing typical components of a HAWT [3].

5. Power generation and transmission systems

Once electricity is generated, it must be prepared for transmission to the grid.

- Power electronics/inverter: Equipment used to convert the variable frequency electrical current from the generator into the stable alternating current (AC) with the voltage and frequency required by the power grid.
- Transformer (step-up transformer): Increases the low voltage of the electricity generated by the turbine to a higher voltage suitable for long-distance transmission over power lines.

1.3 Principles of wind power

1.3.1 The mean-wind velocity

Wind power potential for a given site can be characterized using the mean-wind velocity $\bar{V}_w$ (m/s), which is based on observations over a sufficiently long duration of time. The mean-wind velocity is given by

$$\bar{V}_w = \frac{\sum_i^n V_i}{n} \quad (1)$$

where $\sum_i^n V_i$ is the sum of all velocity measurements and n is the number of the measurements. The sizing of wind energy conversion systems (WECS), would, however, by wholly underpredicted if it were rated at the average wind velocity. A more reasonable approximation would be based on the mean-energy velocity $\bar{V}_E$ (m/s), given by Eq. (2)

$$\bar{V}_E = \left(\frac{\sum_i^n V_i^3}{n} \right)^{1/3} \quad (2)$$

1.3.2 The power available in the incoming wind stream

The total available power of a wind stream $P_{t,w}$ (W) is given by

$$P_{t,w} = \dot{KE}_{in} \quad (3)$$

or,

$$P_{t,w} = \dot{m}_{w,in}\, ke_{in} \quad (4)$$

where $V_{w,in}$ is the incoming wind velocity (m/s) and ke_{in} is the incoming specific kinetic energy (J/kg) of that wind stream, given by

$$ke_{in} = \frac{1}{2} V_{w,in}^2 \quad (5)$$

Using Eq. (5) in Eq. (4), gives

$$P_{t,w} = \frac{1}{2}\, \dot{m}_{w,in} V_{w,in}^2 \quad (6)$$

Using the continuity equation, the mass flow rate of the incoming wind stream $\dot{m}_{w,in}$ (kg/s) is given by

$$\dot{m}_w = \rho_{w,in}\, V_{w,in}\, A \quad (7)$$

where $\rho_{w,in}$ is the density of the incoming wind (kg/m^3) and A is the cross-sectional area (m^2) of the incoming wind stream (later will be referred to as the swept area of the turbine rotor), as shown in **Figure 6**.

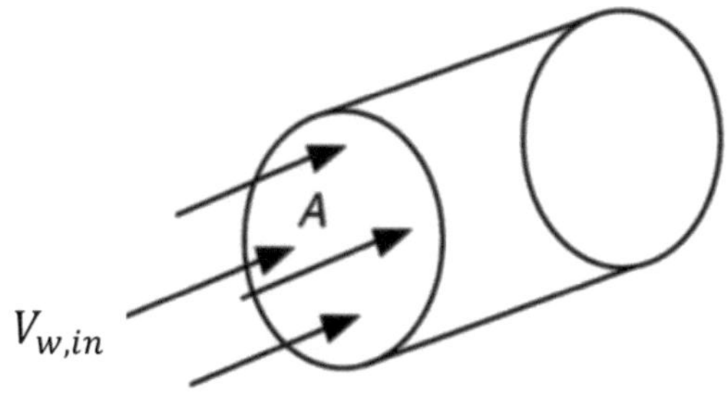

Figure 6.
The stream envelope of an incoming wind.

For a horizontal-axis wind turbine (HAWT), the power output varies linearly with the rotor-swept area, A. This area, defined in terms of the rotor diameter D (m), is given by the following expression:

$$A = \frac{\pi}{4} D^2 \tag{8}$$

Substituting Eq. (7) in Eq. (6) and simplifying, gives

$$P_{t,w} = \frac{1}{2} \rho_{w,in} A \, V_{w,in}^3 \tag{9}$$

For simplicity, the wind (air) density $\rho_{w,in}$ varies with the pressure p (kPa) and temperature T (K) according to the ideal-gas law given by

$$\rho_{w,in} = \frac{p}{RT} \tag{10}$$

where R is air gas constant $(0.287\ kJ/kg.K)$. It should be noted that the air density varies with the site elevation Z (m), and is given by (valid up to 6000 m of a site elevation above sea level) [5].

$$\rho = \rho_{ref} - \left(1.194\, x\, 10^{-4}\, Z\right) \tag{11}$$

The temperature T (K) of the wind varies with elevation Z (m) and is given by

$$T = 288.5 - \frac{19.83\, Z}{3048} \tag{12}$$

The pressure in the international standard atmosphere can be approximated (valid up to a site elevation of 5000 m) by [1]

$$p = 101.29 - 0.011837\, Z + 4.793\, x\, 10^{-7}\, Z^2 \tag{13}$$

A practical parameter known as the wind power density (W/m^2) defined as the wind power divided by the area, is given by

$$\frac{P_{t,w}}{A} = \frac{1}{2} \rho_{w,in}\, V_{w,in}^3 \tag{14}$$

1.3.3 The producible wind power

Depending on the conversion efficiency of a given WECS, part of the available power in the incoming wind can be utilized and produced. This aspect will be discussed in the next section.

2. Modeling and performance aspects of HAWT wind turbine

2.1 The 1-D steady-state wind energy conversion formulation

Consider a wind turbine with a rotor of diameter D and swept area A and with very thin blade m-n (representing an ideal rotor blade), as shown in **Figure** 7. Now, consider the region bounded between (i-m) as shown in **Figure 8** with a control volume (CV) as an energy system.

Application of energy balance for CV_{i-m}, considering steady-state operation of the wind turbine rotor, gives

$$\dot{E}_{in} = \dot{E}_{out} \tag{15}$$

Neglecting potential energy change, the energy balance yields

$$\dot{m}_{w,i}\left(h_{w,i} + \frac{V_{w,i}^2}{2}\right) = \dot{m}_{w,m}\left(h_{w,m} + \frac{V_{w,m}^2}{2}\right) \tag{16}$$

where h_w *(kJ/kg)* is the specific enthalpy of the wind. Since steady-state mass balance states that

$$\dot{m}_{w,i} = \dot{m}_{w,m} \tag{17}$$

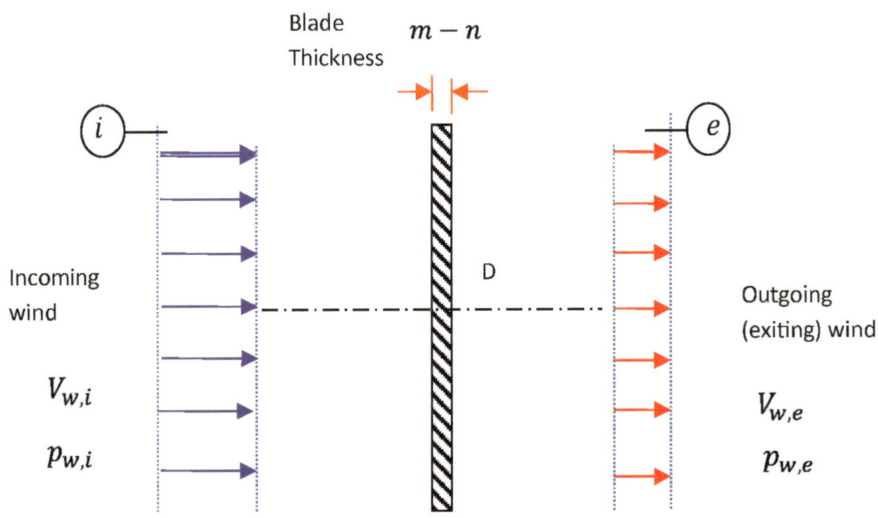

Figure 7.
1-D energy analysis for a very thin blade of an ideal HAWT.

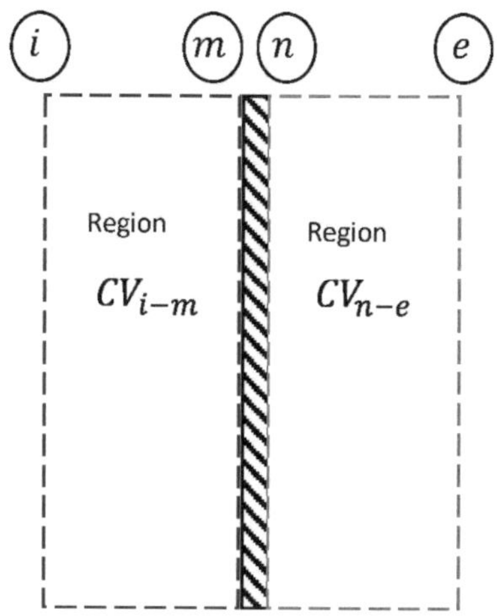

Figure 8.
Bounded regions for the 1-D energy balance analysis.

and recognizing that

$$h = u + pv \tag{18}$$

where *u (kJ/kg)* is the specific internal energy and *v* (m^3/kg) is the specific volume, Eq. (16) yields

$$\left(u_{w,i} + p_{w,i}v_{w,i}\right) + \frac{V_{w,i}^2}{2} = \left(u_{w,m} + p_{w,m}v_{w,m}\right) + \frac{V_{w,m}^2}{2} \tag{19}$$

and since the temperature change of the wind (air) across CV_{i-m} is negligible such that $T_{w,i} = T_{w,m}$ then $u_{w,i} = u_{w,m}$, Eq. (19) yields

$$p_{w,i}v_{w,i} + \frac{V_{w,i}^2}{2} = p_{w,m}v_{w,m} + \frac{V_{w,m}^2}{2} \tag{20}$$

Since, the change in v_w is negligible, such that $v_{w,i} = v_{w,m}$, and replacing v_w by $\left(\frac{1}{\rho_w}\right)$ in Eq. (20), gives

$$p_{w,i} + \rho_w\frac{V_{w,i}^2}{2} = p_{w,m} + \rho_w\frac{V_{w,m}^2}{2} \tag{21}$$

Similarly, considering the region CV_{n-e} and applying energy balance, one can show

$$p_{w,e} + \rho_w\frac{V_{w,e}^2}{2} = p_{w,n} + \rho_w\frac{V_{w,n}^2}{2} \tag{22}$$

Now, since the wind velocity across the turbine blade decreases across the thickness due to the fact that kinetic energy is being converted to mechanical shaft work during the turbine operation within that region, this leads to $V_{w,i} > V_{w,m}$ and $V_{w,n} > V_{w,e}$ such that $p_{w,m} > p_{w,i}$ and $p_{w,n} > p_{w,e}$. Subtracting Eq. (21) from Eq. (22), gives

$$p_{w,m} - p_{w,n} = \left(p_{w,i} + \rho_w \frac{V_{w,i}^2 - V_{w,m}^2}{2}\right) - \left(p_{w,e} + \rho_w \frac{V_{w,e}^2 - V_{w,n}^2}{2}\right) \quad (23)$$

However, in the upstream and downstream the wind pressure is atmospheric so that $p_{w,i} = p_{w,e}$ and realizing that the velocity of the wind across the very-thin bladed turbine is almost constant $V_{w,m} = V_{w,n} = V_T$, Eq. (23) simplifies to

$$p_{w,m} - p_{w,n} = \rho_w \left(\frac{V_{w,i}^2 - V_{w,e}^2}{2}\right) \quad (24)$$

The pressure and velocity profiles across the turbine's rotor are shown in **Figure 9**. Recall, the force in the direction of the 1-D wind stream (i.e., axial force) perpendicular to the swept area A, can be expressed as

$$F_x = \left(p_{w,m} - p_{w,n}\right)A \quad (25)$$

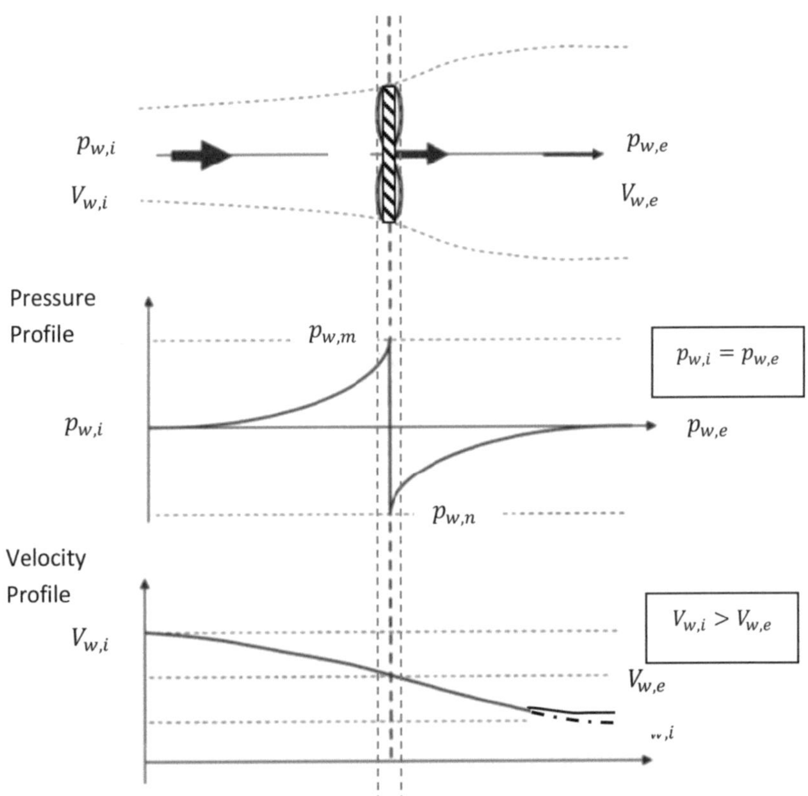

Figure 9.
Pressure and velocity distributions of the wind flow from upstream to downstream of the wind turbine's rotor.

Substitution of Eq. (24) in Eq. (25), gives

$$F_x = \rho_w\, A \left(\frac{V_{w,i}^2 - V_{w,e}^2}{2} \right) \quad (26)$$

Also, recall that the axial force is because of the rate of change of linear momentum, given by

$$F_x = \dot{m}_w (V_{w,i} - V_{w,e}) \quad (27)$$

Using Eq. (7) in Eq. (27) gives

$$F_x = \rho_w\, V_T\, A\, (V_{w,i} - V_{w,e}) \quad (28)$$

Equating Eq. (26) with Eq. (28) yields

$$\rho_w\, A \left(\frac{V_{w,i}^2 - V_{w,e}^2}{2} \right) = \rho_w\, V_T\, A\, (V_{w,i} - V_{w,e}) \quad (29)$$

Simplifying, gives

$$V_T = \frac{1}{2}\, (V_{w,i} + V_{w,e}) \quad (30)$$

2.2 The maximum power generation by a wind turbine – *The Betz limit*

Now, consider the entire system as a control volume CV_{i-e}, application of energy balance, gives

$$\dot{E}_i = \dot{W}_T + \dot{E}_e \quad (31)$$

Applying similar energy analysis, yields

$$\dot{W}_T = \dot{m} \left(\frac{V_{w,i}^2 - V_{w,e}^2}{2} \right) \quad (32)$$

Or,

$$\dot{W}_T = \frac{1}{2}(\rho_w\, V_T\, A)\, \left(V_{w,i}^2 - V_{w,e}^2\right) \quad (33)$$

Substituting Eq. (30) in Eq. (33), gives

$$\dot{W}_T = P_T = \frac{1}{4}(\rho_w\, A)\, (V_{w,i} + V_{w,e})\, \left(V_{w,i}^2 - V_{w,e}^2\right) \quad (34)$$

The maximum power produced by the turbine can be obtained by differentiating P_T in Eq. (33) with respect to $V_{w,e}$ and setting the derivative equal to zero, that is

$$\frac{dP_T}{dV_{w,e}} = 0 \quad (35)$$

In doing so, yields

$$3V_{w,e}^2 + 2V_{w,i}V_{w,e} - V_{w,i}^2 = 0 \tag{36}$$

For a positive $V_{w,e}$ to yield $(V_{w,e})_{opt}$,

$$(V_{w,e})_{opt} = \frac{1}{3}V_{w,i} \tag{37}$$

Substituting back in Eq. (34), gives

$$P_{T,max} = \frac{8}{27}\rho_w \, A \, V_{w,i}{}^3 \tag{38}$$

Defining the maximum theoretical (ideal) turbine efficiency (also known as the maximum turbine power coefficient $C_{p,max}$, Betz limit) is given by

$$\eta_{T,max} = C_{p,max} = \frac{P_{T,max}}{P_{t,w}} = \frac{16}{27} \tag{39}$$

The local velocity experienced by a wind turbine blade is a function of both the inflow angle and the radial distance from the hub. Due to the blade's long geometry, this velocity varies significantly along the span, and the blades are therefore twisted. Empirical data indicate that the maximum efficiency is strongly dependent on what is known as blade-to-wind speed ratio, Ω, given by

$$\Omega = \frac{\omega\left(\frac{D}{2}\right)}{V_{w,i}} \tag{40}$$

Ω reaches its maximum value of approximately 0.6 only when the tip speed ratio—the ratio of the blade's tip speed to the wind velocity—is between 7 and 8. It drops rapidly when this ratio falls below approximately 2.0, as shown in **Figure 10**.

3. Illustrative numerical examples and case studies

3.1 Example 1

It is required to evaluate the viability of installing a horizontal-axis wind turbine of a renewable wind energy system (WECS) on a site for a community near Professor's Lake, Ontario (Latitude: 44° N). The wind turbine of the WECS is capable of converting approximately 60% of its maximum theoretical conversion efficiency has a rotor blade diameter of 110 m. The site's annual average wind speed is observed to be 7.0 m/s. The turbine needs to be steadily operated at a rotational speed of 35 RPM. It is required to estimate the following quantities relevant to the viability of installing this WECS on the site. The site has ambient conditions of atmospheric pressure (=*101.325 kPa*) and an average temperature of $10^{\circ}C$.

Assumption: Air can be assumed as an ideal gas with *R* = *0.287 kJ/kg.K.*

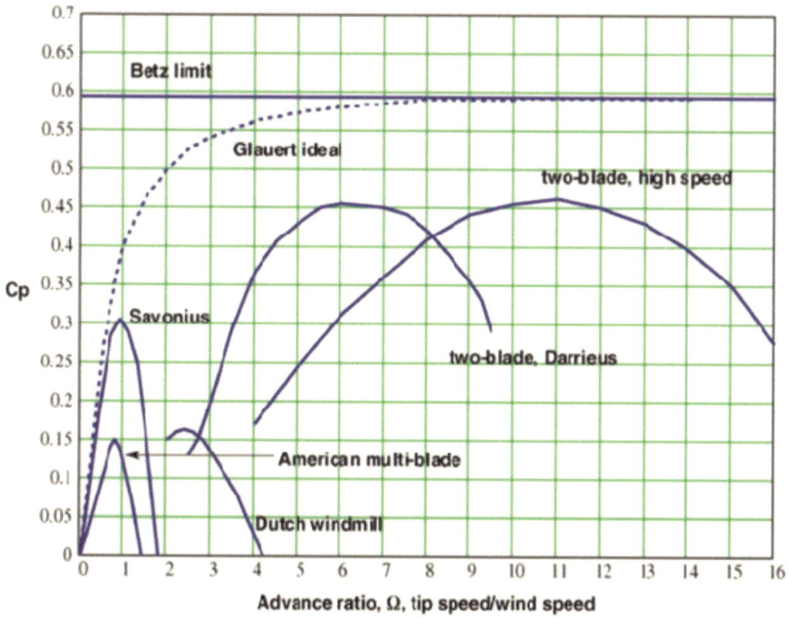

Figure 10.
Power coefficient vs. advance ratio for various wind turbine configurations [2].

a. The available total wind power density (W/m^2) in the incoming wind.

b. The available total wind power (in kW).

c. The maximum producible power density from the wind turbine.

d. The optimum wind exit velocity from the turbine and the corresponding wind velocity across the turbine blade and the mass flow rate of air (kg/s).

e. The actual producible power density from the turbine.

f. The maximum producible power from the turbine.

g. The actual turbine producible power.

h. The torque ($kN.m$) if the turbine to be operated at the maximum efficiency.

i. The maximum axial thrust (kN) acting on the turbine bearing.

j. The producible useful power from the WECS to match an average DC load if the WECS has the following components' efficiencies (**Table 1**).

Analysis:

a. Recall Eq. (14), gives

$$\frac{P_{t,w}}{A} = \frac{1}{2}\rho_{w,in}\, V_{w,in}^3 \tag{41}$$

WECS component	Average efficiency (%)
Battery system	85
Charge controller	90
Rectifier	95

Table 1.
Components' efficiencies for the WECS in Example 1.

The wind density can be estimated using Eq. (10)

$$\rho_{w,in} = \frac{1.01325\ x\ 10^5}{287\ (10 + 273)} = 1.247\ kg/m^3 \tag{42}$$

Substituting in Eq. (41), gives

$$\frac{P_{t,w}}{A} = \frac{1}{2}(1.247)(7)^3 = 213.9\ W/m^2 \tag{43}$$

b. Using Eqs. (8) and (43), give

$$P_{t,w} = 213.9\ (9503.32) = 2032.8\ kW$$

c. Recall Eq. (39),

$$\frac{P_{T,max}}{A} = \eta_{T,max}\left(\frac{P_{t,w}}{A}\right)$$

$$\frac{P_{T,max}}{A} = \frac{16}{27}\ (213.9) = 126.8\ W/m^2$$

d. Recall Eq. (37),

$$V_{w,e} = \frac{1}{3} V_{w,i}$$

$$V_{w,e} = \frac{1}{3}(7) = 2.33\ m/s$$

Using Eq. (30)

$$V_T = \frac{1}{2}(7 + 2.33) = 4.67\ m/s$$

Recall Eqs. (7) and (8),

$$\dot{m}_w = \rho_{w,in}\ V_T\ A = 1.247\ (4.67)(9503.32) = 55342.5\ kg/s$$

e. Recall Eq. (39),

$$\frac{P_{T,act}}{A} = \eta_{T,act}\left(\frac{P_{T,max}}{A}\right) = \eta_{T,act}\eta_{T,max}\left(\frac{P_{t,w}}{A}\right)$$

Substituting, gives

$$\frac{P_{T,act}}{A} = (0.6)\left(\frac{16}{27}\right)(213.9) = 76.05\ W/m^2$$

f. Recall Eq. (8),

$$P_{T,max} = \frac{126.8\ (9503.32)}{1000} = 1205.0\ kW$$

g. Recall Eq. (8),

$$P_{T,act} = \frac{76.05\ (9503.32)}{1000} = 722.7\ kW$$

h. Introducing,

$$P_{T,max} = \mathrm{T}_{max}\ \omega = \mathrm{T}_{max}\left(\frac{2\pi N}{60}\right)$$

Solving for T_{max}, gives

$$\mathrm{T}_{max} = 1205.0\left(\frac{60}{2\pi(35)}\right) = 328.8\ kN.m$$

i. Recall Eq. (28),

$$F_x = \rho_w\ V_T\ A\ (V_{w,i} - V_{w,e}).$$

The maximum axial force occurs when $V_{w,e} = \frac{1}{3}V_{w,i}$, substituting back, gives

$$F_{max} = \rho_w\ V_T\ A\left(V_{w,i} - \frac{1}{3}V_{w,i}\right)$$

Substituting, yields

$$F_{max} = \frac{(1.247)(4.67)(9503.32)}{1000}\left(\frac{2}{3}\right)(7) = 258.27\ kN$$

j. Introducing,

$$P_{WECS} = \eta_{WECS}\, P_{t,w}$$

$$P_{WECS} = \eta_{T,act}\, \eta_{max}\eta_{Rec}\, \eta_{cc}\eta_{B}\, P_{t,w}$$

Substituting, gives

$$P_{WECS} = (0.6)\left(\frac{16}{27}\right)(0.95)(0.90)(0.85)(2032.8) = 525.3\ kW$$

3.2 Example 2

It A *15 m/s* wind at 101.3 *kPa* and 20°*C* enters a two-bladed wind turbine with a diameter of 15 *m*. Show step-by step analysis and estimate the following:

a. The power of the incoming wind.

b. The theoretical max power that could be extracted.

c. The efficient producible turbine power.

d. The speed in RPM required for part (c).

e. The torque associated with part (c).

Analysis:

a. Recall, Eq. (10)

$$\rho_{w,in} = \frac{1.01325\ x\ 10^5}{287\ (10 + 273)} = 1.205\ kg/m^3$$

Using Eq. (14), gives

$$P_{t,w} = \frac{1}{2}(1.205)\left(\frac{\pi}{4}\right)(15)^2(15)^3 = 359.34\ kW$$

b. Recall Eq. (39),

$$P_{T,max} = \eta_{T,max}\ P_{t,w} = \left(\frac{16}{27}\right)(359.34) = 212.94\ kW$$

c. For a two-bladed HAWT, with an advance ratio $\Omega = 11.0$, $\eta_T = C_{p,T} = 0.46$ (**Figure 10**).

Substituting,

$$P_T = \eta_T \, P_{T,max} = (0.46)(212.94) = 165.30 \; kW$$

d. Recall Eq. (40),

$$\Omega = \frac{\omega\left(\frac{D}{2}\right)}{V_{w,i}} = 11 = \frac{\omega\left(\frac{15}{2}\right)}{15}$$

Solving for ω, yields

$$\omega = 22 \; rad/s$$

Thus,

$$N = \frac{60(22)}{2\pi} = 210.1 \; RPM$$

e. Recall Eq. (40),

$$\mathrm{T} = \frac{P_{T,act}}{\omega} = \frac{165.30}{22} = 7.513 \; kN.m$$

3.3 Example 3

Consider the rotor blades of a HAWT which turns a large shaft at low speed of rotation, **Figure 4**. The wind turbine produces 2.5 MW of electrical power. The Nacelle has a diameter D_N of 3 m and length L_N of 6 m. The energy conversion efficiencies of the gearbox and generator are 93 and 95%, respectively. The wind (air) temperature is $25^{\circ}C$ with an estimated average convection heat transfer coefficient of $35 \; W/m^2.^{\circ}C$. Determine the rate of heat transfer dissipated from the Nacelle, $\dot{Q}_N$ (kW) under steady-state operating conditions.

Analysis:

Under steady-state operating conditions, applying energy balance over the nacelle, Eq. (15) yields

$$\dot{W}_{mech,in} = \dot{W}_T + \dot{Q}_N \tag{44}$$

Rearranging, gives

$$\dot{Q}_N = \dot{W}_{mech,in} - \dot{W}_T \tag{45}$$

Since the energy conversion efficiency for the gearbox η_{GB} is given by

$$\eta_{GB} = \frac{\dot{W}_{GB}}{\dot{W}_{mech,in}} \tag{46}$$

Rearranging, yields

$$\dot{W}_{mech,in} = \frac{\dot{W}_{GB}}{\eta_{GB}} \tag{47}$$

Since the energy conversion efficiency for the electric generator η_{EG} is given by

$$\eta_{EG} = \frac{\dot{W}_{T}}{\dot{W}_{GB}} \tag{48}$$

Rearranging, yields

$$\dot{W}_{GB} = \frac{\dot{W}_{T}}{\eta_{EG}} \tag{49}$$

Substituting Eq. (49) in Eq. (47), gives

$$\dot{W}_{mech,in} = \frac{\dot{W}_{T}}{\eta_{EG}\,\eta_{GB}} \tag{50}$$

And substituting Eq. (50) back in Eq. (45), yields

$$\dot{Q}_{N} = \dot{W}_{T}\left(\frac{1}{\eta_{EG}\,\eta_{GB}} - 1\right) \tag{51}$$

Substituting the numerical values, gives

$$\dot{Q}_{N} = 2500\left(\frac{1}{0.95x0.95} - 1\right) = 330\ kW \tag{52}$$

4. Conclusion

As a proven and sustainable resource for power generation, wind energy is a critical component of the renewable landscape. This chapter covered the fundamental principles of wind turbines, with a specific focus on modeling the power generation of horizontal-axis wind turbines (HAWTs). A detailed discussion of the governing formulations was supported by numerical examples and illustrative case studies that demonstrate their applications. It was found through detailed 1-D energy formulations that the maximum wind turbine conversion efficiency is approximately 59.3% which confirms the Betz limit.

Author details

Basel I. Abed Ismail
Department of Mechanical and Mechatronics Engineering, Lakehead University, Thunder Bay, Ontario, Canada

*Address all correspondence to: baseliai@gmail.com

References

[1] Khaligh A, Onar OC. Solar, Wind, and Ocean Energy Conversion Systems. NW, USA: Taylor & Francis Group, LLC, CRC Press; 2010

[2] Manwell JF, McGowan JG, Rogers AL. Wind Energy Explained: Theory, Design and Application. 2nd ed. The Atrium, Southern Gate, West Sussex, UK: John Wiley & Sons Ltd.; 2009

[3] Hodge BK. Alternative Energy Systems and Applications. 2nd ed. New Jersey, USA: John Wiley & Sons; 2017

[4] Sumathi S, Kumar LA. Solar PV and Wind Energy Conversion Systems. Switzerland: Springer International Publishing; 2015

[5] Patel MR. Wind and Solar Power Systems. 2nd ed. NW, USA: Taylor & Francis Group, LLC, CRC Press; 2006

Chapter 2

Enhancing the Utilization of Wind Resources through Artificial Intelligence Techniques

Monica Borunda

Abstract

The integration of wind energy into power systems provides many benefits and also involves numerous challenges. Wind's intermittency across temporal and spatial scales is one of the biggest drawbacks of this source and requires accurate forecasting strategies to ensure grid stability, economic and environmental benefits, and effective operational planning. This chapter sightsees the use of artificial intelligence (AI) techniques to enhance wind resource utilization, focusing on wind power forecasting across four key time horizons: very short-term, short-term, medium-term, and long-term. Each forecast horizon is helpful for addressing various operational issues, from real-time control and ramp-rate management to strategic investment planning. This chapter provides an overview of the most common AI models, highlighting their strengths, limitations, and applicability in relation to the forecast horizon, compared to conventional approaches. The aim of this work is to obtain a deeper understanding of how AI can support a more efficient and reliable wind energy future by an adequate treatment of wind resource variability through good predictions of the resource availability at different time scales. The chapter concludes with a discussion of emerging trends, such as hybrid models, model interpretability, and the integration of forecasting tools to achieve the best use of wind energy.

Keywords: wind power, wind resource, forecasting, artificial intelligence, machine learning

1. Introduction

Nowadays, efforts from all parties worldwide are focusing on accomplishing the global transition towards reaching Net Zero 2050. Wind energy has positioned itself as one of the key sources of energy to achieve this, reaching a global installed wind capacity of 900 GW by 2023. Moreover, climate targets, policy incentives, and cost competitiveness continue to point towards its expansion [1, 2]. However, the intermittent nature of this source poses significant challenges for grid integration, operational planning, and the electrical market.

Wind is an inherently intermittent and stochastic resource, subject to fluctuations across time and space, from seconds to years and meters to kilometers. This variability directly affects wind power output, as it depends on the wind speed cubed, and thus, it can experience rapid changes, known as ramps, or prolonged periods of underperformance [3, 4]. As a result, wind power systems must deal with increased uncertainty in generation, which complicates balancing supply and demand, maintaining frequency stability, and ensuring grid reliability [5]. Moreover, the inability to accurately forecast these fluctuations can lead to a deficiency in covering the demand and activating the energy reserve, and even more, in the need to rely on backup conventional generation, reducing the economic and environmental benefits of wind energy [6]. Addressing this variability requires forecasting systems that are not only accurate but also adaptive, robust, and capable of operating under uncertainty [7].

In the last years, Artificial Intelligence (AI) has stood out for its ability to model complex systems, uncover hidden patterns in large datasets, and improve decision-making processes in dynamic environments [8]. In the energy sector, AI has been successfully applied to support the integration of renewable energy sources and optimize the operation of hybrid energy systems [9, 10]. Its strength lies in the flexibility of machine learning algorithms to deal with nonlinear, noisy, and high-dimensional data, which characterize typical datasets of intermittent resources.

AI has also enhanced the use of renewable energies, for example, by assessing solar and wind resources that, due to their intermittency, are challenging to manage. The assessment of intermittent renewable resources, like wind and solar energy, goes from resource power forecasting, analyzing their impacts on the system, and incorporating control strategies to compensate for the power fluctuations by introducing energy storage or other alternatives. In the context of wind power forecasting, AI methods have improved or even overcome traditional statistical techniques in terms of accuracy and performance [11–13]. These methods enable not only accurate short-term predictions but also reasonable long-term estimations that incorporate uncertainty and climatic variability. As the scale and complexity of wind energy systems increase, AI offers a scalable, data-driven solution to forecast generation, manage variability, and support reliable system operation.

Depending on the forecast application, the forecast horizon is defined, and depending on the time scale, the model characteristics, the input data, the spatial scale, the evaluation metrics, and the challenges change [14]. The time scale runs across four main time horizons: very short-term (seconds to minutes), short-term (15 minutes to 6 hours), medium-term (6 hours to 7 days), and long-term (weeks to years). By addressing the specific challenges, strengths, advantages, and disadvantages of various techniques at the different forecast horizons, this chapter presents an overview of the current status of AI applied to wind power forecasting.

This chapter does not attempt to provide a survey or review of all works published in wind power forecasting using AI. Instead, it attempts to present the basics behind wind power forecasting, as well as its usefulness, along with an overview of some of the applications of AI techniques to wind power forecasting, citing some examples, in such a way that the reader is presented a panorama to understand the root of the problem and the scope that AI can reach to enhance the utilization of wind energy.

The chapter is organized as follows. Section 2 describes the challenges encountered when forecasting in wind power systems. Section 3 presents the main forecast horizons, categorized by their usefulness in wind power generation. Finally, the chapter is concluded in Section 4, and future perspectives are presented.

2. Forecasting challenges in wind power systems

2.1 Sources of uncertainty in wind power forecasting

Wind power forecasting is subject to multiple factors of uncertainty due to the complex and dynamic behavior of the atmosphere, the nonlinear behavior of wind turbines, and the limitations of accurate data.

At the meteorological level, wind is governed by complex interactions across multiple spatial and temporal scales. Atmospheric turbulence, rapid mesoscale changes, convective phenomena, and weather fluctuations introduce a high intermittency in wind speed and direction. This intermittent behavior is stronger in regions with complex orography or abrupt land-sea transitions, where local effects can increase the wind profile intermittence.

A crucial factor that magnifies forecasting error in wind power prediction is that wind speed variations propagate through nonlinear transformations in the turbine's power curve, amplifying errors at the output level. Theoretically, the power output of a wind turbine is proportional to the cube of the wind speed, as shown in Eq. (1),

$$P_{wind} \propto \frac{1}{2}\,\rho\,A\,v^3, \tag{1}$$

where P_{wind} is the wind power, ρ is the air density, A is the area swept by the blades of the wind turbine, and v is the wind speed [15]. Moreover, Betz's law states that the maximum power, P_{max}, a wind turbine can extract from the wind is [16, 17].

$$P_{max} \sim 59.3\% P_{wind}. \tag{2}$$

Therefore, the electrical power output of a wind turbine is proportional to the cube of the wind speed within the operational range, which lies between the cut-in and rated speed as shown in **Figure 1**.

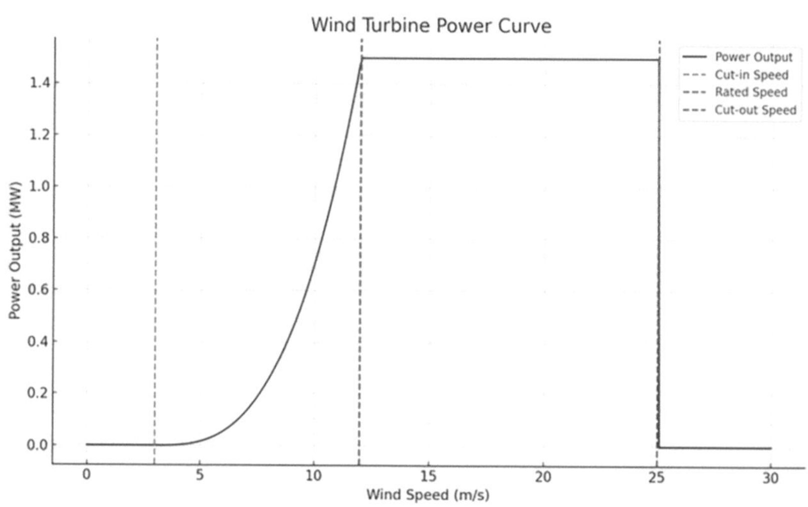

Figure 1.
Power curve of a wind turbine, which describes the electrical power output of a wind turbine as a function of the wind speed. The cut-in speed corresponds to the wind speed at which the wind turbine generates power with a cubic dependence on the wind speed. The rated speed is the wind speed at which the wind turbine generates the nominal power. The cut-out speed corresponds to the wind speed at which the wind turbine ceases to operate.

Thus, even minor errors in forecasting wind speed can result in disproportionately large deviations in power generation; for example, a ±10% error in the wind speed implies a ±30% error in the wind power forecasting. Thus, this cubic dependence of the generated wind power on the wind speed amplifies the sensitivity of forecasting models to uncertainties in meteorological inputs. These uncertainties are caused by measurement errors resulting from sensor calibration or misalignment. Additionally, wind speed varies with height as

$$\frac{v_2}{v_1} = \left[\frac{z_1}{z_2}\right]^{\alpha}, \tag{3}$$

where v_1 and v_2 are the wind speeds at heights z_1 and z_2 and α is the Hellmann exponent and depends upon the shape and location of the terrain and the stability of the air [18]. This fact is significant for wind speed forecasting at the lowest and highest points of the area swept by the wind turbine, in particular for large wind turbines. Moreover, even though there are instruments that measure wind speeds at different heights, such as light detection and ranging (LiDAR) and sonic detection and ranging (SoDAR), their resolution is affected by fog, turbulence, and weather fluctuations [19].

Wind speed values at turbine hub height can be obtained by other means, such as through numerical weather prediction (NWP) models [20]. However, these methods rely on the numerical solution of physics equations describing the dynamics of the atmosphere, which require boundary conditions to be solved. The boundary conditions correspond to local measurements, and often they are insufficient; thus, the results reduce their predictive reliability at turbine hub height [21]. Moreover, wake interactions, orographic effects, and local thermal gradients further complicate the accurate modeling of wind inflow. These compounded uncertainties represent a critical barrier for both deterministic and probabilistic forecasting models [22–24].

2.2 Spatio-temporal resolution considerations

The accuracy of wind power forecasting is strongly influenced by the spatio-temporal resolution of the input data and the forecasting model. Spatial and temporal resolution directly influence the predictive skill of AI-based forecasting models. Models trained on high-resolution historical datasets tend to perform better for site-specific forecasting, whereas low-resolution data may lead to increased bias and reduced accuracy. Spatial resolution refers to the granularity at which wind speed and direction, as well as other meteorological variables, are used in the model. In contrast, temporal resolution defines the frequency of data sampling and forecast updates. Both dimensions are critical for capturing the inherent variability of wind resources and for meeting the operational requirements of power systems.

From a spatial perspective, higher resolution allows the model to capture local topographic effects, roughness changes, and microscale wind conditions that can significantly affect turbine performance. This is important in complex terrains or offshore environments, where local wind events, such as wake effects or coastal breezes, can lead to deviations from large-scale predictions. For instance, a model operating at 1 km resolution may provide significantly better forecasts for a specific wind farm than a 10 km resolution NWP model, as the former sees local wind gradients more accurately [25, 26]. Increasing spatial resolution demands many accurate weather stations and good computational resources [27]. Currently, there are efforts focused on improving spatial resolution with novel techniques [28].

From a temporal perspective, the required resolution depends on the forecast horizon. In any forecast process, regardless of the horizon, a large database of many years of wind speed data is required. The bigger the dataset, the more accurate the forecast is likely to be. Very short-term forecasts generally require wind speed data obtained at intervals of seconds to minutes. On the other hand, short-term forecasts require a database with time intervals from minutes to hours, and, in particular, day-ahead forecasting can require data intervals of 5, 10, 15, 30, or 60 minutes. Finally, medium-term and long-term forecasting requests hourly data [29]. **Figure 1** summarizes this information (**Figure 2**).

The temporal resolution also interacts with spatial resolution, as higher temporal resolution is more effective when coupled with adequate spatial granularity [30]. Currently, efforts are focused on improving temporal resolution using novel techniques [31–33].

An important consideration is the compromise between resolution and computational cost. Increasing spatial and temporal resolution improves the quality of wind data and thus is more helpful for the appropriate treatment of the wind resource; however, it also increases the computational workload, particularly in hybrid forecasting frameworks that integrate numerical weather prediction with machine learning models [34–36]. This balance becomes critical for real-time forecasting applications, where the rapid processing becomes crucial [37, 38].

In summary, achieving an optimal spatio-temporal resolution is essential for a good performance of wind power forecasting systems. It requires careful consideration of the site characteristics and measurement equipment, forecast horizon, operational requirements, and available computational resources. Thus, these needs impose requirements on model design. Hybrid approaches, which combine physical, statistical, and/or machine learning models across spatial and temporal scales, have appeared as a promising strategy to address these challenges [39].

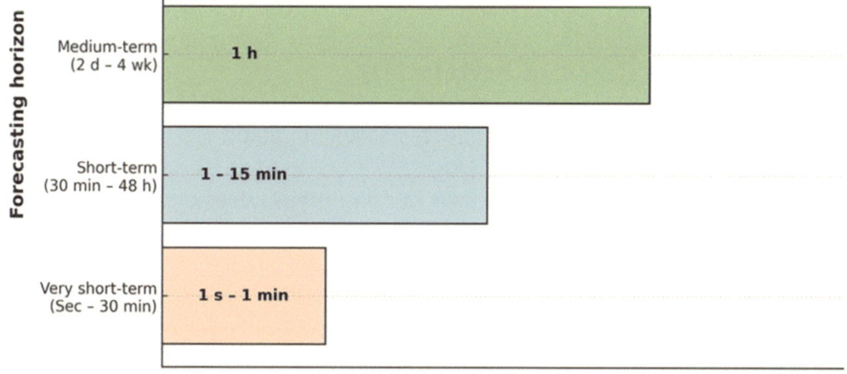

Figure 2.
Recommended temporal data frequency by forecast horizon.

2.3 Data quality and availability constraints

Forecasting approaches, particularly those based on artificial intelligence, are highly sensitive to the quality and availability of historical datasets. Nowadays there are many sources to collect data, the main ones being: meteorological measuring towers, SCADA (Supervisory Control and Data Acquisition) systems, remote sensing technologies (e.g., LiDAR and SoDAR), and satellite equipment. However, these sources frequently present problems such as missing values, calibration errors, and spatial-temporal misalignments. Additionally, confidentially agreements are required to access to high-resolution operational data from wind farm operators, being very difficult to obtain. The lack of standardized data preprocessing practices, such as handling ramp events or sensor drift, also reduces model reproducibility and generalization.

The performance of wind power forecasting models—whether based on NWP, statistical approaches, or advanced machine learning techniques—depends fundamentally on the quality and availability of the input data. High-quality, representative datasets enable accurate model training and validation, while poor-quality or incomplete datasets introduce biases, increase uncertainty, and limit generalization capabilities.

Data quality in wind energy forecasting can be evaluated in several aspects:

- The accuracy is the difference between measured or simulated data and real conditions. Instrument calibration errors, sensor drift, and weather conditions directly influence accuracy in data [40, 41].

- The resolution refers to the temporal and spatial granularity of the data. Very short and short-term forecasts require high-resolution measurements, between 1 and 10 minutes, whereas medium and long-term forecasts require hourly or daily data [23, 42–45].

- The completeness of the data refers to the proportion of missing values within the dataset due to operational failures in sensors and communication devices or maintenance events [46, 47]. Missing values can reduce the dataset's representativeness.

- The consistency of datasets refers to uniform measurement protocols, formats, and units, and it is fundamental for multi-source integration [48].

- The readiness of data is the state and quality of the data and is crucial, in particular for very short-term forecasts [49].

- Likewise, there are sources that may decrease data quality, such as:

- Measurement errors caused by sensor misalignment, anemometer wear, and blasé shadow effects on nacelle-mounted instruments.

- Data gaps in wind speed data due to communication failures, maintenance stops, and bad weather conditions.

- Environmental factors, such as icing and turbulence, can bias wind speed measurements [50–55].

- Drawbacks in SCADA systems which may hide short-term variability needed for high-resolution forecasting [56, 57].

- Systematic biases present in atmospheric datasets from reanalysis, such as MERRA-2 and ERA5, may not be favorable for accurate wind power forecasting [58–60].

- Also, there may be data availability constraints like

- Proprietary restrictions due to confidentiality policies of SCADA owners [61].

- Limited coverage of meteorological stations in remote or offshore sites [62, 63].

- Licensing software and equipment costs may not be accessible for everyone or any research project.

- Insufficient data because there may not be enough historical data to capture interannual variability or climate change effects.

Thus, numerous issues are involved in the crucial step of selecting the appropriate data for accurate wind power forecasting. In particular, AI-based forecasting models are data-hungry, requiring large, diverse, and representative datasets to achieve high performance and avoid drawbacks such as overfitting, or reducing its ability to model critical events [64].

To attenuate these problems, some actions can be taken as follows [65]:

- Use of filters to detect anomalies and sensor faults in real-time.

- Use of gap-filling techniques to reconstruct missing values.

- Alignment of model outputs with observed measurements.

- Data augmentation and multi-source data fusion to improve the dataset.

- Integration of distributed sensor networks and community weather stations to increase spatial coverage.

- Implementation of open data formats and protocols for wind speed and wind turbine data to facilitate interoperability.

2.4 Operational, economic, and regulatory implications

Wind power forecasting systems not only are limited to the technical issues of developing the appropriate methodologies for a successful prediction but also have implications in the operational, economic, and regulatory aspects of the power system. Accurate forecasting is essential to maintain system reliability, optimize market participation, and comply with regulatory frameworks.

Operationally, wind power forecasting is essential to ensuring the balance between supply and demand in power systems. System operators rely on day-ahead and intra-day forecasts to schedule generation resources and manage reserves [66, 67]. Mistaken forecasts imply an increase in reserve requirements, which raise operational costs, and ramp events which may destabilize the grid [68]. Moreover, in real-time conditions, wind power forecasting is crucial for frequency regulation and voltage support.

Economically, wind power forecasting is essential in liberalized electricity markets where inaccurate forecasts can lead to suboptimal bidding strategies, increased imbalance penalties, and reduced revenues for wind power producers [69–71]. Therefore, enhanced forecasting abilities are essential not only for operational reliability but also for supporting market participation, reducing integration costs, and enhancing the utilization of wind resources in power systems [72]. Moreover, poor forecasts increase balancing costs and position wind resources in disadvantage with respect to conventional fuels. Likewise, reliable long-term forecasts are essential for investment decisions providing assessment for wind farm projects. A small reduction in wind power forecasting errors can imply annual savings of millions of dollars for large power systems [73].

Also, regulatory frameworks increasingly demand the use of accurate wind power forecasts. For instance, grid codes, corresponding to the technical standards that govern the connection and operation of generation assets, require system operators to ensure balance, stability and reliability, and thus make wind power day-ahead forecasting an essential regulatory obligation for the secure integration of wind energy into power systems [6, 74–76]. Wind power forecasting is also relevant for national energy strategies to facilitate higher penetration of renewables and meet international agreements on clean energy production [77, 78].

Hence, wind power forecasting, at different time scales, enhances the utilization of wind resources from the operational, economic, and regulatory aspects and boosts its integration into power systems.

3. AI-based wind power forecasting by time horizon

Wind power forecasting is a complex task with different challenges depending on the temporal horizon of the prediction, which may range from seconds to years. As mentioned before, physical, statistical, artificial intelligent, and hybrid methods are the main approaches for wind power forecasting at different horizons [7]. Physical models, such as ensemble predictions and numerical weather models, are the most commonly used [79]. The most widely used non-intelligent statistical models are primarily time series analysis methods, including Persistence, Autoregressive (AR), Autoregressive Moving Average (ARMA), Autoregressive Integrated Moving Average (ARIMA), and Seasonal Autoregressive Integrated Moving Average (SARIMA) models, which are among the most popular ones [11, 80]. Most of these methods are suitable for very short and short-term forecasting. In contrast, AI-based models are more recent and are able to deal with any forecast horizon due to their ability to learn patterns across multiple temporal scales, as demostrated in the following [7, 81–83]. Finally, hybrid methods combine several methods, taking advantage of the benefits of each one and achieving, most of the time, excellent results [7, 11, 84, 85].

It is important to note that the accuracy of the forecast is quantified with a metric. The common metrics used in wind power forecasting are shown in **Table 1**.

Acronym	Name	Physical meaning	Formula
RMSE	Root mean square error	Average magnitude of error	$RMSE = \sqrt{\frac{1}{N}\sum_{i=1}^{N}(x_i - \hat{x}_i)^2}$
MAE	Mean absolute error	Level of absolute error	$MAE = \frac{1}{N}\sum_{i=1}^{N}\lvert x_i - \hat{x}_i\rvert$
MAPE	Mean absolute percentage error	Percentage error relative to the actual value	$MAPE = \frac{1}{N}\sum_{i=1}^{N}\left\lvert\frac{x_i - \hat{x}_i}{x_i}\right\rvert \cdot 100\%$
MSE	Mean squared error	Prediction error	$MSE = \frac{1}{N}\sum_{i=1}^{N}(x_i - \hat{x}_i)^2$
nRMSE	Normalized root mean squared error	Normalized average magnitude of error	$nRMSE = \frac{RMSE}{x_{max} - x_{min}}$
nMAE	Normalized mean absolute error	Normalized level of absolute error	$nMAE = \frac{MAE}{x_{max} - x_{min}}$
R2	R-square or coefficient of determination	Correlation between predicted and actual value	$R2 = 1 - \frac{\sum_{i=1}^{N}(x_i - \hat{x}_i)^2}{\sum_{i=1}^{N}(x_i - \overline{x})^2}$

Table 1.
Main metrics for evaluating the accuracy in wind power forecasting.

where x_i and $\hat{x}_i$ are the actual and predicted N values of the variable respectively, x_{max} and x_{min} are the maximum and minimum values, and $\overline{x}$ is the average of the actual values [86]. The next subsections describe the different forecast horizons, their applications, and the AI models used to afford them.

3.1 Very short-term forecasting

Very short-term wind power forecasting refers to the prediction of wind power at time horizons ranging from a few seconds to a few minutes ahead. Since the forecast horizon is the smallest possible, it is also referred to as real-time wind power forecasting. It is essential for real-time grid control and stability, as well as frequency and voltage regulation, spinning and non-spinning reserves, and black start capability. Conventional methods using time series have a very good performance in this forecasting. However, AI methods have four big advantages over conventional methods suitable for these problems, which are the ability to (a) process high-volume data, (b) detect patterns, (c) deal with nonlinearity, and (d) deal with complex data.

Reported studies have shown that ANNs have been widely used for wind power forecasting. For instance, Zha and Jing [87] use them for 15-minute-ahead forecasting in China, evaluating their results with RMSE, MAE, and MAPE, finding excellent accuracy. Other AI tools, such as random forest, gradient boosting, and extreme gradient boosting, are applied for 10-minute-ahead wind power forecasting [88]. They obtained excellent results reflected in the calculated metrics, MAE, MSE, RMSE, and R2, acquiring, for example, an R-squared of up to 0.999. Five minutes ahead wind power forecasting is performed in Australia using more sophisticated machine learning tools such as convolutional and long-short-term memory neural networks together with a support vector machine in [89]. Accuracy performance is reported in terms of RMSE and MAE, achieving a bit more than 20%. Likewise, the same analysis was performed using convolutional and recurrent neural networks for different wind farms in Australia, reporting their performance with MAE and RMSE, showing that accuracy depends on the site, managing to obtain up to 1.5% in MAE [90]. Wind speed forecasting for China with a 15-minute ahead horizon is performed using a hybrid method that combines statistics and a neural network together with a genetic algorithm [91]. Performance is calculated with different metrics, achieving errors of less than 10%. A comparison study has been performed for 10-minute-ahead wind power forecasting with data from Morocco using support vector machine, deep neural network, and recurrent neural network [92]. They measure the accuracy obtained by the three approaches, calculating MSE, MAE, MAPE, and R2, finding out that the recurrent neural network performs the best with an R2 of 0.9.

Their approaches also involve data decomposition and feature extraction to reduce their complexity and then use an AI technique to determine behaviors or parameters, such as neural networks – backpropagation, convolutional, and recurrent, support vector machines, data mining algorithms, and genetic algorithms [93]. Likewise, a hybrid model based on ensemble empirical mode decomposition, whale optimization, and an Elman NN was constructed for ultra-short-term wind power, obtaining a higher accuracy in their results compared with other models [94].

Hybrid methods of probabilistic wind speed forecasting have also been developed, where prediction intervals are defined using probabilistic criteria and then, wind speed mean values every 20 minutes are forecasted with an LSTM NN in order to obtain the probability distribution of predicted wind speeds [95]. Also, wind speed, wind direction, and wind power forecasting 1 hour ahead have been performed using Bayesian NNs at 80 and 90% confidence levels [96].

Moreover, explainable artificial intelligence (XAI) has also been employed to identify the critical elements in their 15-minute-ahead wind power forecasting hybrid model composed of LSTM NN, GRU, and ARIME algorithms [97].

Conventional methods' performance for very short-term forecasting is very good for very short-term wind power forecasting, in particular, the statistical ones. However, let us keep in mind that AI is ideal for dealing with complex structures; thus, AI techniques, such as RNNs and LSTM NNs, can be applied to capture temporal dependencies in turbulent wind patterns. Other approaches, such as CNNs and hybrid models, can be applied in wind power forecasting models employing signal decomposition. Moreover, hybrid approaches combining probability and AI techniques are also useful for wind power forecasting, providing a quantified uncertainty.

Table 2 summarizes the main applications and approaches concerning intelligent wind power forecasting at the very short term.

Intelligent very-short-term wind power forecasting models also present limitations, the most significant one being the need for a large dataset. Artificial intelligence core lies in training the models, and for that, the more available data, the better the training and the better the performance. Another limitation of intelligent models lies in the fact of overfitting, which refers to the fact that the model fits the training data of a complex model very well, by memorizing it, but is unable to predict real data. Finally, most of the models may be able to perform very well at the site used for their construction, but their performance cannot be generalized to different sites.

In the following subsection, wind power forecasting at a longer time horizon is discussed.

3.2 Short-term forecasting

Short-term wind power forecasting refers to time horizons ranging from several hours to one day ahead. This forecast is useful for unit commitment, economic dispatch, day-ahead bidding in the electrical market, and scheduling of energy storage; thus, it is essential for the security of supply and the economy of power systems. Indeed, errors in this time horizon are reflected in economic and energy losses, reserve misallocation, and inefficiencies in the grid operation.

Time ahead	Application	Approach to forecasting	AI techniques
From minutes to a few hours ahead.	• Frequency regulation; • Real-time pitch and yaw adjustments; • Reserve allocation; • Grid code; • Electricity market.	Use time series of wind speed data input to find sequential dependencies and patterns.	• DNN • RNN • LSTM NN
		Use time series of wind power and other data as input to capture sequential dependencies and patterns.	• DNN • RNN • LSTM NN • Random forest • Gradient boosting • Extreme gradient boosting
		Use data decomposition to forecast behavior of parameters.	Hybrid methods with • DNN • RNN • CNN • SVM • DMA • GA
		Perform probabilistic forecasting.	Hybrid methods with • LSTM NN • Bayesian NN
		Explain the crucial factors in the forecasting process.	• -XAI

Table 2.
Main facts about very short term wind power forecasting with AI techniques.

Numerical weather prediction models perform very well at this time horizon. These models involve physical modeling of the atmosphere's dynamics and require boundary conditions at the site of interest for hours-ahead or day-ahead wind speed forecasting. The boundary conditions are given by measured data from meteorological stations or, in the worst case, interpolated data. The major challenge in these cases is that the topography of the site or sudden local meteorological phenomena alter the reasonable accuracy of the prediction, which, given the dependence of wind power on the wind speed as specified in Eq. 1, increases the error in the power prediction. Statistical models for one-day-ahead forecasting have been developed, but the results strongly depend on the seasonality of the data [14]. Artificial intelligence has been extensively applied to contribute to this field [80].

The popular ANNs have been extensively used for short-term wind speed and wind power forecasting. Data from an SCADA system has been used as an input for wind speed forecasting [98]. Another example is an ANN that has been used for day-ahead forecasting, finding the best network architecture with sensitivity analysis and comparing results with NWP [99]. Moreover, by optimizing ANN parameters, good accuracy has been found in [100]. Moreover, in [101], a performance comparison for three-day-ahead forecasting with ANN and ANN with optimized weights with two different intelligent methods, Firefly Algorithm and Particle Swarm Optimization, is made, showing that the latter outperforms the former, and the model can be scalable. Some of the most used AI techniques, such as gradient boosting, random forest, and transformer NNs, have been used for short-term wind power forecasting and compared, turning out that the best accuracy is achieved with gradient boosting with an

R2 of 0.98, as long as the methodology is data-driven [102]. Models involving RNNs have also been designed with successful performance [103]. LSTM NNs have also been used for day-ahead wind power forecasting to extract the short-term temporal dependency of the input variables, which are wind power and weather variables time-series, finding an nRMSE of about 10% [104].

CNNs have also been utilized to attend day-ahead wind power forecasting as shown in [105], where an RMSE of 0.629 is attained. Moreover, an LSTM-based model is constructed with excellent results [106]. Additionally, different complex models have been developed and compared. For instance, three different types of combinations have been tested for day-ahead wind power forecasting, which are CNN with Bi-LSTM NNs, CNN with biGRUs, and LSTM with GRU, with the former being the one with the best performance, achieving an MAE of 0.41 [107]. Also, wind speed forecasting has been promisingly performed with Bi-directional LSTM, accomplishing an R2 of 0.87 [108]. Different approaches use CNN. In particular, a model built with CNN together with a Bi-LSTM is able to enhance wind power forecasting by 28% during power fluctuation periods [109].

Probability density forecasting of wind power has also been performed at this time horizon. As an example, wind power forecasting is done using different conditional quantiles by means of a quantile regression NN [110]. A different intelligent probabilistic approach is employed by combining a Weather Research and Forecasting simulation, optimization, and clustering treatment in [111]. Wind speed and wind power forecasting at a horizon of 5 hours ahead have been performed with Dynamic Bayesian Networks, showing it as a promising probabilistic intelligent forecasting tool [112]. Moreover, probability density forecasting of wind power has also been attained with more novel AI technology in [113], where a Transformer Network has been used, combined with other techniques.

Hybrid models have also been developed to address this forecast horizon. For example, wind power forecasting with GRU NN and ARIMA has been performed for a couple of hours ahead with meteorological and wind farm data from a location in Pakistan, obtaining an R2 of 0.89 [114]. Hybrid models can become very complex. For example, for day-ahead forecasting, a CNN model is employed to analyze the spatial correlations between wind power and NWP data, then a BiGRU model studies the temporal dependencies in the time-series data, and finally, they merge the results with a multi-head self-attention mechanism to allocate greater weights to the most influential elements of the forecasting model [115]. In a different hybrid model, wind power data are decomposed into high and low frequencies, and CNNs are used to obtain spatial features which will serve as inputs to an LSTM NN [116].

Table 3 summarizes the main characteristics and approaches of intelligent wind power forecasting in the short term.

Even though intelligent short-term wind power forecasting models have their strengths, such as the ability to deal with huge datasets, non-linearity, and complex and large temporal and spatial patterns, there are also weak points, such as the imperative need for large datasets, high computational requirements, high computational cost, risk of overfitting, among others. And more importantly, if the intelligent model manages to forecast with excellent accuracy, it is unlikely that the same result will be obtained at different locations. However, this is a field where much still needs to be done, and efforts continue in this direction.

Medium-term wind power forecasting is discussed in the following subsection.

Time ahead	Application	Approach to forecasting	AI techniques
From hours to a day ahead.	• Unit commitment; • Economic dispatch; • Day-ahead bidding in the electrical market.	Use time series of wind speed data and other variables as inputs to find sequential dependencies and patterns.	• ANN
		Use time series of wind power and other data as inputs to capture sequential dependencies and patterns.	• RNN • Gradient boosting • Random forest • LSTM NN • Bi-LSTM NN • Bi-GRU • Transformer NN • CNN
		Perform probabilistic forecasting.	Hybrid models with • QRNNs • clustering • Dynamic Bayesian Networks • QRNNs Quantile Regresion
		Analyze spatial and temporal patterns.	Hybrid models with • CNN • GRU • Bi-GRU • LSTM NN

Table 3.
Main facts about short-term wind power forecasting with AI techniques.

3.3 Medium-term forecasting

Medium-term forecasting refers to predictions performed from 2 to 17 days ahead, up to some weeks ahead. These predictions are useful for preventive maintenance and component replacement of the wind power systems. It is also useful for improving grid reliability by providing a future scenario of wind power generation, thereby facilitating planning of energy reserves and cross-border interconnection planning in the event of low wind events. Additionally, this medium-term forecasting is useful to plan energy balancing strategies and energy trading. Physical and statistical methods have also been used to present solutions in this time horizon. Physical models such as the Global Forecast System operated by NOAA provide deterministic and probabilistic weather data up to 2 weeks in advance. Moreover, there are other methods based on NWP, such as ECMWF, ENS, GEFS, that have been used for wind speed forecasting from days to weeks ahead. Since these are global models, sometimes they are used with statistical downscaling, though linear regression or other techniques, and using datasets from reanalysis, like ERA 5, to tune the forecast to local conditions at the wind turbine or farm. Additionally, there are statistical models, such as the Weather Regime Forecasting, to infer the weather regime and provide weather values up to a few weeks [117].

Additionally, there are forecasting models that incorporate AI techniques. Monthly mean wind speed forecasting has been done with ECMWF, enhancing the accuracy by means of a calibration performed with decision trees, random forest, and light gradient boosting [118]. For instance, ANN, like the Multilayer Perceptron and backpropagation, have been successful in medium to long-term wind power forecasting [119, 120]

Time ahead	Application	Approach to forecasting	AI techniques
From days to weeks ahead.	• Grid reliability; • Preventive maintenance; • Component replacement; • Planning energy reserve; • Planning cross-border interconnection; • Planning of energy balance strategies; • Planning of energy trading.	Use time series of wind speed data and other variables as inputs from NWP models to improve accuracy.	• Decision trees • Random forest • Gradient boosting
		Use time series of wind power and other data as inputs to capture sequential dependencies and patterns.	• ANN • Random forest • Gradient boosting • Multilayer perceptron NN • Backpropagation NN • k nearest neighbors regression
		Perform probabilistic forecasting.	• Bayesian Networks
		Analyze spatial and temporal patterns.	Hybrid models with • Transformers NN

Table 4.
Main facts about medium-term wind power forecasting with AI techniques.

respectively. Also, a gradient boosting-based approach has been introduced for short- and medium-term wind power forecasting, achieving good accuracy [121]. Models with different techniques, ANN, random forest, extreme gradient boosting, k-nearest neighbors' regression, and multilayer perceptron, have forecasted wind power at short, medium, and long-term, finding out that random forest provides more accurate forecasts for medium and long-term [122]. A probabilistic intelligent approach has also been explored by using Bayesian networks for wind power forecasting a month ahead to assess the penetration of small-scale wind power in the residential sector [123].

Hybrid methods for wind speed and wind power forecasting at this timescale have also been presented. Also, novel AI technology has been employed to improve classical methods. NWP results have been corrected by means of a transformer neural network, obtaining good results [124].

Table 4 recapitulates the main characteristics and approaches of intelligent wind power forecasting at the medium term.

Medium-term prediction is more complex than short-term prediction, and this fact is reflected in a smaller number of publications in the literature. As before, the use of AI techniques has its advantages, such as dealing with non-linearity, handling long sequences, recognizing patterns, and managing uncertainties in forecasting. However, the same disadvantages persist, such as the need for large datasets, the risk of overfitting, and the complexity of the models. Nevertheless, it opens an alternative to conventional methods with more and more innovative tools as time goes by. This provides an opportunity to use artificial intelligence techniques, traditional or the most recent, or even hybrid models, which, together with the creativity of researchers, can provide promising results applicable to various locations.

Finally, long-term wind power forecasting is discussed in the next subsection.

3.4 Long-term forecasting

Long-term wind power forecasting has the longest prediction interval, ranging from months to years. Long-term wind power fluctuations are affected by wind speed variations at daily, seasonal, or annual scales, and thus, the probability of obtaining less accurate results is higher. Additionally, local weather conditions can influence long-term wind behavior, making the problem even more complex. Nevertheless, it has important technical applications, such as operation management, optimal operation, dispatch planning, site selection, resource assessment, and system expansion. Additionally, it has important economic applications, like information for financial investments and bankable documentation.

Statistical approaches have been used to study the persistent behavior of the wind, as well as physical methods, with the aid of downscaling techniques of NWP. Many of the approaches use wind resource seasonality to help, in one way or another, to enhance the accuracy obtained in the forecasted values [125–127]. AI-based models have also been developed [128, 129].

ANNs have been used extensively in this forecast horizon. Monthly wind power forecasting using an ANN in Sri Lanka has been performed, and various error measures have been computed, including R^2 and MSE, with R^2 values ranging from 0.97 to 1 and MSE values ranging from 0.02 to 0.00 [130]. Also, ANNs have been used to predict hourly values of wind speed for the next year by recognizing long-term patterns and then adjusting the prediction [131]. Likewise, there are approaches with LSTM NNs, which, by considering regional characteristics, have improved the wind-resource prediction at medium and long-term [132]. Another approach uses Support Vector Regression, with polynomial and radial basis functions, to estimate the parameters of the Weibull distribution characteristic of the wind resource at the site at long-term [133]. Also, a multivariate data set has been used as input for different AI models, LSTM NN, GRU, CNN, and CNN-LSTM, for wind power forecasting at three different horizons: 6 months, 1 year, and 5 years. The results show that the GRU method achieves higher accuracy [134].

A probabilistic intelligent approach has been proposed using clustering techniques to find statistical seasonality and define a wind resource typical year, from which, by means of the probability distribution function, the annual energy for the coming year is estimated [135].

Explainable AI has also been employed by introducing an X-double LSTM, which incorporates XAI and LSTM NN, to predict wind production a year ahead. The results

Time ahead	Application	Approach to forecasting	AI techniques
From weeks to months or years ahead.	• Operation management; • Dispatch planning; • Site selection; • Resource assessment; • Wind system project planning; • System expansion planning; • Financial information.	Use time series of wind speed data and other variables as inputs from NWP models to improve accuracy.	• ANN • LSTM NN
		• Use time series of wind power and other data as inputs to capture sequential dependencies and patterns.	• ANN • LSTM NN • Bi-LSTM NN • GRU • CNN • Transformer NN
		Perform probabilistic forecasting.	• Bayesian n etworks
		Explain the crucial factors in the forecasting process.	• XAI: X-double LSTM NN

Table 5.
Main facts about long-term wind power forecasting with AI techniques.

have been compared with the ones obtained with LSTM, Bi-LSTM, GRU, transformer model, and ARIMA models, finding an increase in accuracy, aiding operational decision-making performance [136].

Table 5 summarizes the main characteristics and approaches of intelligent wind power forecasting at the medium term.

As mentioned before, the longer the forecast horizon, the greater the complexity of the forecasting process and the less accurate the results become. At this forecast horizon, it is more suitable to use a probabilistic approach that includes the intermittent nature of the wind resource rather than a deterministic approach. Thus, various hybrid approaches, combining probabilistic methods with AI, have emerged in recent years. Moreover, the problem of a model obtaining good forecasts for one site but not being able to generalize to others persists across all forecast horizons.

4. Conclusions and future perspectives

One of the main actions to improve the use of wind resources lies in better understanding this intermittent resource at all timescales, in order to take full advantage of it. This is achieved through a good forecast of the wind resource at different horizon forecasts. Enhancing the utilization of wind resources through AI techniques not only relies on increasing the forecasting accuracy but also on obtaining an integrated decision-support system with reliable and interpretable forecasts that can be directly used by system operators, market actors, and policymakers to use this resource as a strong pillar of power systems.

So far, all prediction paradigms have advantages and limitations, as shown in **Table 6**, which indicate that the best path towards more accurate and reliable forecasting systems is through the hybridization of AI methods with or without non-intelligent ones. So far, there is no single method that surpasses the others.

To conclude, AI should not be considered as a replacement for the non-intelligent methods, but as an additional method to create synergy to overcome the limitations of the other methodologies. Nowadays, AI offers the possibility to provide accurate and interpretable results by means of explainable AI, which enhances the utilization of

Approach	Advantages	Limitations
Physical models	They are based on a physical model, and thus they are physically consistent and interpretable.	They need to solve complex equations with a complex input database and deal with fine-scale variability.
Numerical weather prediction models	They provide global and clear forecasts across different time scales.	They are computationally expensive and are sensitive to local conditions.
Statistical methods	They are relatively simple, fast, and adaptable to specific sites.	They may underperform in extreme events, and they may not be extrapolated.
Artificial Intelligence techniques	They deal with nonlinearity, multivariable, and spatiotemporal dependencies. They can integrate different types of data from different types of sources.	They need a large amount of data and computational requirements. There is a high risk of overfitting. Most of the methods perform like black boxes.

Table 6.
Advantages and limitations of the different approaches for wind power forecasting.

wind resources in the operational and market scope of power systems. AI opens the door to data from different technologies that may incorporate crucial information for the forecasting results. Finally, probabilistic intelligent models may be the best way to deal with medium and long-term, as well as regional forecasting, which provides useful information to harness the exploitation of this resource.

Acknowledgements

MB thanks SECIHTI for her national research position with ID 71557 and CENIDET-TECNM for its hospitality.

Conflict of interest

The authors declare no conflict of interest.

Acronyms and abbreviations

AI	artificial intelligence
ANN	artificial neural network
Bi-LSTM NN	bidirectional long short term neural network
CNN	convolutional neural network
GRU	gated recurrent unit
LSTM NN	long short-term memory neural network
NN	neural network
NWP	numerical weather prediction
RNN	recurrent neural network
XAI	explainable artificial intelligence
X-double LSTM NN	explainable bidirectional long short term memory neural network

Author details

Monica Borunda
SECIHTI – Tecnológico Nacional de México – Centro Nacional de Investigación y Desarrollo Tecnológico, Morelos, Mexico

*Address all correspondence to: monica.bp@cenidet.tecnm.mx

References

[1] GWEC. Global Wind Report 2023. Brussels, Belgium: Global Wind Energy Council; 2023. Available from: https://gwec.net/global-wind-report-2023/

[2] IEA (2024), Renewables 2023, IEA, Paris. Available from: https://www.iea.org/reports/renewables-2023, Licence: CC BY 4.0

[3] Notton G, Nivet ML, Voyant C, Paoli C, Darras C, Motte F, et al. Intermittent and stochastic character of renewable energy sources: Consequences, cost of intermittence and benefit of forecasting. Renewable and Sustainable Energy Reviews. 2018;**87**: 96-105. DOI: 10.1016/j.rser.2018.02.007

[4] Holttinen H et al. Design and Operation of Power Systems with Large Amounts of Wind Power. Finland: IEA Wind Task; 2009, Final Report. p. 25

[5] Ahmed SD, Al-Ismail FSM, Shafiullah M, Al-Sulaiman FA, El-Amin IM. Grid integration challenges of wind energy: A review. IEEE Access. 2020;**8**: 10857-10878. DOI: 10.1109/ACCESS.2020.2964896

[6] Loza B, Minchala LI, Ochoa-Correa D, Martinez S. Grid-friendly integration of wind energy: A review of power forecasting and frequency control techniques. Sustainability. 2024;**16**(21): 9535. DOI: 10.3390/su16219535

[7] Tsai WC, Hong CM, Tu CS, Lin WM, Chen CH. A review of modern wind power generation forecasting technologies. Sustainability. 2023;**15**(14): 10757. DOI: 10.3390/su151410757

[8] Mukhamediev RI et al. Review of artificial intelligence and machine learning technologies: Classification, restrictions, opportunities and challenges. Mathematics. 2022;**10**(15): 2552. DOI: 10.3390/math10152552

[9] Olawade DB et al. Artificial intelligence potential for net zero sustainability: Current evidence and prospects. Next Sustainability. 2024;**4**: 100041. DOI: 10.1016/j.nxsust.2024.100041

[10] Rolnick D. Tackling climate change with machine learning. ACM Computing Surveys. 2022;**55**(2):1-96. DOI: 10.1145/348512

[11] Hanifi S, Liu X, Lin Z, Lotfian S. A critical review of wind power forecasting methods – Past, present and future. Energies. 2020;**13**(15):3764. DOI: 10.3390/en13153764

[12] Deng X, Shao H, Hu C, Jiang D, Jiang Y. Wind power forecasting methods based on deep learning: A survey. Computer Modeling in Engineering and Sciences. 2020;**122**:1. DOI: 10.32604/cmes.2020.08768

[13] Karaman OA. Prediction of wind power with machine learning models. Applied Sciences. 2023;**13**(20):11455. DOI: 10.3390/app132011455

[14] Li F, Wang H, Wang D, Liu D, Sun K. A review of wind power prediction methods based on multi-time scales. Energies. 2025;**18**:1713. DOI: 10.3390/en18071713

[15] Wagner HJ, Mathur J. Introduction to Wind Energy Systems. Basics, Technology and Operation. Heidelberg New York Dordrecht London: Springer; 2018. DOI: 10.1007/978-3-319-68804-6

[16] Betz A. Das Maximum der theoretisch möglichen Ausnutzung des Windes durch Windermotoren. Zeitschrift für das gesamte Turbinenwesen. 1920;**26**:307-309

[17] Betz A. Wind-energie und ihre Ausnutzung durch Windmühlen. Göttingen: Vandenhoeck & Ruprecht; 1926

[18] Kaltschmitt M, Streicher W, Wiese A. Renewable Energy: Technology, Economics and Environment. Springer Science & Business Media; 2007. DOI: 10.1007/3-540-70949-5

[19] Lang S, Lidar MK. Sodar measurements of wind speed and direction in upland terrain for wind energy purposes. Remote Sensing. 2011; **3**(9):1871-1902. DOI: 10.3390/rs3091871

[20] Aggarwal R, Kumar R. A comprehensive review of numerical weather prediction models. International Journal of Computer Applications. 2013; **74**(18):44-48. DOI: 10.5120/12989-0246

[21] Yang M, Jiang Y, Che J, Han Z, Lv Q. Short-term forecasting of wind power based on error traceability and numerical weather prediction wind speed correction. Electronica. 2024;**13**(8):1559. DOI: 10.3390/electronics13081559

[22] Yan J, Möhrlen C, Göcmen T, Kelly M, Wessel A, Giebel G. Uncovering wind power forecasting uncertainty sources and their propagation through the whole modelling chain. Renewable and Sustainable Energy Reviews. 2022;**165**: 112519. DOI: 10.1016/j.rser.2022.112519

[23] Gayathry V, Deepa K, Sangeetha SVT, Porselvi T, Ramprabhakar J, Gowtham N. Uncertainty analysis of different forecast models for wind speed forecasting. Renewable Energy. 2025;**241**:122285. DOI: 10.1016/j.renene.2024.122285

[24] Mararakanye N, Dalton A, Bekker B. Incorporating spatial and temporal correlations to improve aggregation of decentralized day-ahead wind power forecasts. IEEE Access. 2022;**10**: 116182-116195. DOI: 10.1109/ACCESS.2022.3219602

[25] Giovannini L, Antonacci G, Zardi D, Laiti L, Panziera L. Sensitivity of simulated wind speed to spatial resolution over complex terrain. Energy Procedia. 2014;**59**:323-239. DOI: 10.1016/j.egypro.2014.10.384

[26] Pryor SC, Nikulin G, Jones C. Influence of spatial resolution on regional climate model derived wind climates. Journal of Geophysical Research Atmospheres. 2012;**17**:D03117. DOI: 10.1029/2011JD016822

[27] Mbugua JM, Hiraga Y. Recent advances in long-term wind-speed and -power forecasting: A review. Climate. 2025;**13**(8):155. DOI: 10.3390/cli13080155

[28] Kumar A, Islam T, Ma J, Kashiyama T, Sekimoto Y, Mattmann C. WindSR: Improving spatial resolution of satellite wind speed through super-resolution. IEEE Access. 2023;**11**: 69486-69494. DOI: 10.1109/ACCESS.2023.3292966

[29] Tsai WC, Hong CH, Tu CS, Lin WM, Chen CH. A review of modern wind power generation forecasting technologies. Sustainability. 2023;**15**(14): 10757. DOI: 10.3390/su151410757

[30] Benmoufok EF, Warder SC, Zhu E, Bhaskaran B, Staffell I, Piggott MD. Improving wind power modelling through granular spatial and temporal bias correction of reanalysis data. Energy. 2024;**313**:133759. DOI: 10.1016/j.energy.2024.133759

[31] Omoyele O, Hoffmann M, Koivisto M, Larrañeta M, Weinand JM,

Linssen J, et al. Increasing the resolution of solar and wind time series for energy system modeling: A review. Renewable and Sustainable Energy Reviews. 2024; **189**(part B):113792. DOI: 10.1016/j.rser.2023.113792

[32] Chen F, Vinsen K, Filoche A. Spatial-temporal approach for gridded wind forecasting across Southwest Western Australia. IEEE Access. 2024;**12**:185905-185917. DOI: 10.1109/ACCESS.2024.3485096

[33] Wilhelm P, Eggert M, Hornig J, Oertel S. High spatial and temporal resolution bistatic wind Lidar. Applied Sciences. 2021;**11**(16):7602. DOI: 10.3390/app11167602

[34] Merizzi F, Asperti A, Colamonaco S. Wind speed super-resolution and validation: From ERA5 to CERRA via diffusion models. Neural Computing and Applications. 2024;**36**: 21899-21921

[35] Hadjipetrou S, Kyriakidis P. High-resolution wind speed estimates for the eastern Mediterranean Basin: A statistical comparison against coastal meteorological observations. Wind. 2024;**4**(4):311-341. DOI: 10.3390/wind4040016

[36] Benton BN, Buster G, Pinchuk P, Glaws A, King RN, Maclaurin G, et al. Super-resolution for renewable energy resource data with wind from reanalysis data and application to Ukraine. Energies. 2025;**18**(14):3769. DOI: 10.3390/en18143769

[37] Joseph L, Deo RC, Prasad R, Salcedo-Sanz S, Raj N, Soar J. Renewable Energy. 2023;**204**:39-58. DOI: 10.1016/j.renene.2022.12.123

[38] Landberg L, Giebel G, Nielsen HA, Nielsen T, Madsen H. Short-term prediction an overview. Wind Energy. 2003;**6**:273280. DOI: 10.1002/we.96

[39] Mundotiya P, Mathuria P, Tiwari HP. Comprehensive review on uncertainty in wind power forecasting-models and challenges. In: 2022. 2nd Interenational Conference on Innovative Sustainable Computational Technologies (CISCT). Dehradun, India. 2022. pp. 1-6. DOI: 10.1109/CISCT55310.2022.10046605

[40] Azorin-Molilna C, Asin J, McVicar T, Minola L, López-Moreno JL, Vicente-Serrano SM, et al. Evaluating anemometer drift: A statistical approach to correct biases in wind speed measurement. Atmospheric Research. 2017;**203**:175-188. DOI: 10.1016/j.atmosres.2017.12.010

[41] Spiliotis E, Petropoulos F, Nikolopoulos K. The impact of imperfect weather forecasts on wind power forecasting performance: Evidence from two wind farms in Greece. Energies. 2020;**13**(8):1880. DOI: 10.3390/en13081880

[42] Zhang F, Ren X, Liu Y. A refined wind power forecasting method with high temporal resolution based on light convolutional neural network architecture. Energies. 2024;**17**:8. DOI: 10.3390/en17051183

[43] Yan J, Zhao X, Li K. On temporal resolution selection in time series wind power forecasting. In: 2016 UKACC 11th International Conference on Control, Belfast, UK. 2016. DOI: 10.9/CONTROL.2016.7737573

[44] Liu G, Zhang S, Zhao H, Liu J, Liang G, Zhao J, et al. Super-resolution perception for wind power forecasting by enhancing historical data. Frontiers in Energy Research. 2022;**10**:959333. DOI: 10.3389/fenrg.2022.959333

[45] Resifi S, Al Aawar E, Dasari HP, Jebari H, Hoteit I. A novel deep learning approach for regional high-resolution spatio-temporal wind speed forecasting for energy applications. Energy. 2025; **328**:136356. DOI: 10.1016/j.energy.2025.136356

[46] Tawn R, Browell J, Dinwoodie I. Missing data in wind farm time series: Properties and effect on forecasts. Electric Power Systems Research. 2020; **189**:106640. DOI: 10.1016/j.epsr.2020.106640

[47] Jing B, Pei Y, Qian Z, Wang A, Zhu S, An J. Missing wind speed data reconstruction with improved context encoder network. Energy Reports. 2022; **8**:3386-3394. DOI: 10.1016/j.egyr.2022.02.177

[48] Menezes D, Mendes M, Almeida JA, Farinha T. Wind farm and resource datasets: A comprehensive survey and overview. Energies. 2020;**13**(18):4702. DOI: 10.3390/en13184702

[49] Mou X, Lin W, He Y. Towards a consistent wind data record for the CFOSAT scatterometer. Remote Sensing. 2023;**15**(8):2081. DOI: 10.3390/rs15082081

[50] Atalan YA, Atalan A. Testing the wind energy data based on environmental factors predicted by machine learning with analysis of variance. Applied Sciences. 2025;**15**(1):241. DOI: 10.3390/app15010241

[51] Tang S, Tiantian L, Guo Y, Zhu R, Qu H. Correction if various environmental influences on Doppler wind lidar based on multiple linear regression model. Renewable Energy. 2022;**184**:933-9947. DOI: 10.1016/j.renene.2021.12.018

[52] Bosnjakovic M, Hrkac F, Stoic M, Hradovi I. Environmental impact of wind farms. Environments. 2024;**11**(11):257. DOI: 10.3390/environments11110257

[53] Gonzalez E, Valldecabres L, Seyr H, Melero J. On the effects of environmental conditions on wind turbine performance: An offshore case study. Journal of Physics Conference Series. 2019;**1356**(1):012043

[54] Ozturkoglu O, Ozcelik O, Günel S. Effects of operational and environmental conditions on estimated dynamic characteristics of a large in-service wind turbine. Journal of Vibration Engineering and Technologies. 2024;**12** (Suppl 1):803-824. DOI: 10.1007/s42417-024-01447-4

[55] Pandit R, Infield D, Santos M. Accounting for environmental conditions in data-driven wind turbine power models. IEEE Transactions on Sustainable Energy. 2023;**14, 1**:168-177. DOI: 10.1109/TSTE.2022.3204453

[56] Rautmare S. SCADA system security: Challenges and recommendations. In: Annual IEEE India Conference, Hyderabad, India. 2011. pp. 1-4. DOI: 10.1109/INDCON.2011.6139567

[57] Rustamova S, Rustamov F. Problems of integrating artificial intelligence withh SCADA systems. ScienceRise. 2024;**2**: 106-117. DOI: 10.21303/2313-8416.2024.003691

[58] Qiao Y, Ji D, Shang H, Xu J, Xu R, Shi C. The fusion of ERA5 and MERRA-2 atmospheric temperature profiles with enhanced spatial resolution and accuracy. Remote Sensing. 2023;**15**(14): 3592. DOI: 10.3390/rs15143592

[59] Santos JS, Sakagami Y, Haas R, Passos JC, Machuca NM, Radünz WC, Dias E, Lima MM. Wind speed

evaluation of MERRA-2, ERA-Interim and ERA-5 Reanalysis data at a wind farm located in Brazil. Conference: ISES Solar World Congress 2019/IEA SHC International Conference on Solar Heating and Cooling for Buildings and Industry 2019. International Solar Energy Society Selection. DOI: 10.18086/swc.2019.45.10

[60] Gruber K, Regner P, Wehrle S, Zeyringer M, Schmidt J. Towards global validation of wind power simulations: A multi-country assessment of wind power simulation from MERRA-2 and ERA-5 reanalyses bias-corrected with the global wind atlas. Energy. 2022;**238**:121520

[61] Fahad A, Tari Z, Almalawi A, Goscinski A, Khalil I, Mahmood A. PPFSCADA: Privacy preserving framework for SCADA data publishing. Future Generation Computer Systems. 2017;**37**:496-511. DOI: 10.1016/j.future.2014.03.002

[62] Al-Hinai A, Charabi Y, Kaboli SHA. Offshore wind energy resource assessment across the territory of Oman: A spatial-temporal data analysis. Sustainability. 2021;**13**(5):2862. DOI: 10.3390/su13052862

[63] Rehman S, Irshad K, Ibrahim NI, AlShaikhi A, Mohandes MA. Offshore wind power resource assessment in the Gulf of North Suez. Sustainability. 2023; **15**(21):15257. DOI: 10.3390/su152115257

[64] Zhao E, Sun S, Wang S. New developments in wind energy forecasting with artificial intelligence and big data: A scientometric insight. Data Science and Management. Jun 2022; **5**(2):84-95. DOI: 10.1016/j.dsm.2022.05.002

[65] Liu H, Chen C. Data processing strategies in wind energy forecasting models and applications: A comprehensive review. Applied Energy. 2019;**249**(3):392-408. DOI: 10.1016/j.apenergy.2019.04.188

[66] Soini V. Wind power intermittency and the balancing power market: Evidence from Denmark. Energy Economics. 2021;**100**:105381. DOI: 10.1016/j.eneco.2021.105481

[67] Minhas DM, Khalid RR, Frey G. Load control for supply-demand balancing under renewable energy forecasting. In: 2017. IEEE Second International Conference on DC Microgrids (ICDCM), Nuremburg. Germany: IEEE, Piscataway; 2017. pp. 365-370. DOI: 10.1109/ICDCM.2017.8001071

[68] Goodarzi S, Perera HN, Bunn D. The impact of renewable energy forecast errors on imbalance volumes and electricity spot prices. Energy Policy. 2019;**134**:110827. DOI: 10.1016/j.enpol.2019.06.035

[69] Fatras N, Ma Z, Duan H, Jorgensen BN. A systematic review of electricity market liberalization and its alignment with industrial consumer participation: A comparison between the Nordics and China. Renewable and Sustainable Energy Reviews. 2022;**167**: 112793. DOI: 10.1016/j.rser.2022.112793

[70] Bojnec S, Krizaj A. Electricity markets during the liberalization: The case of a European Union country. Energies. 2021;**14**(14):4317. DOI: 10.3390/en14144317

[71] Ali H, Phoumin H, Suryadi B, Farooque AA, Yaqub R. Assessing ADEAN's liberalized electricity markets: The case of Singapore and the Philippines. Sustainability. 2022;**14**(18): 11307. DOI: 10.3390/su141811307

[72] Zhang B, He G, Du Y, Wen H, Huan X, Xing B, et al. Assessment of the

economic impact of forecasting errors in peer-to peer energy trading. Applied Energy. 2024;**374**:123750. DOI: 10.1016/j.apenergy.2024.123750

[73] Ringel M. Liberalizing European electricity markets: Opportunities and risks for a sustainable power sector. Renewable and Sustainable Energy Reviews. 2003;**7**(6):485-499. DOI: 10.1016/S1364-0321(03)00069-8

[74] Verdejo-Fredes H, García-Muñoz F, Tobar F, Becker C, Olivares M, Zolezzi J, et al. Retail electricity market liberalization: An overview of international experience and effects on the Chilean regulated tariff. Sustainability. 2022;**14**(21):13996. DOI: 10.3390/su142113996

[75] Nomandela S, Mnguni MES, Raji AK. A case study on the validation of renewable energy grid code compliance for a large-scale wind power plant grid-connected mode of operation in real-time simulation. Applied Sciences. 2025;**15**(10):5521. DOI: 10.3390/app15105521

[76] Chen B, Li Z, Li S, Zhao Q, Liu X. A wind power prediction framework for distributed power grids. Energy Engineering. 2024;**121**(5):1291-1307. DOI: 10.32604/ee.2024.046374

[77] Verdone A, Panella M, De Santis E, Rizzi A. A review of solar and wind energy forecasting: From single-site to multi-site paradigm. Applied Energy. 2025;**392**:126016. DOI: 10.1016/j.apenergy.2025.126016

[78] Lenzi V, Ulbig A, Andersson G. Impacts of forecast accuracy on grid integration of renewable energy sources. In: IEEE Grenoble Conference. France: Grenoble; 2013. pp. 1-6. DOI: 10.1109/PTC.2013.6652486

[79] Lange M, Focken U. New developments in wind energy forecasting. In: IEEE Power and Energy Society General Meeting – Conversion and Delivery of Electrical Energy in the 21st Century. Pittsburgh, PA, USA: IEEE; 2008. pp. 1-8. DOI: 10.1109/PES.2008.4596135

[80] Tuncar EA, Saglam S, Oral B. A review of short-term wind power generation forecasting methods in recent technological trends. Energy Reports. 2024;**12**:197-209

[81] Amato F, Guignard F, Walch A, et al. Spatio-temporal estimation of wind speed and wind power using extreme learning machines: Predictions, uncertainty and technical potencial. Stoch Environ Res Risk Assess. 2022;**36**:2049-2069

[82] Yan J, Han X, Wang H, Ge C, Liu Y. An AI-based weather prediction method for wind farms combining global forecast field and wind speed temporal transfer characteristics. Energy. 2025;**329**:136740. DOI: 10.1016/j.energy.2025.136740

[83] Wang X, Guo P, Huang X. A review of wind power forecasting models. Energy Procedia. 2011;**12**:770-778. DOI: 10.1016/j.egypro.2011.10.103

[84] Liu Z, Guo H, Zhang Y, Zuo Z. A comprehensive review of wind power prediction based on machine learning: Models, applications, and challenges. Energies. 2025;**18**(2):350. DOI: 10.3390/en18020350

[85] Soman SS, Zareipour H, Malik O, Mandal P. A review of wind power and wind speed forecasting methods with different time horizons. In: North American Power Symposium, Arlington, TX, USA. 2010. pp. 1-8. DOI: 10.1109/NAPS.2010.5619586

[86] Piotrowski P, Rutyna I, Baczynski D, Kopyt M. Evaluation metrics for wind power forecasts: A comprehensive review and statistical analysis of errors. Energies. 2022;**15**(24):9657. DOI: 10.3390/en15249657

[87] Zha L, Jing DX. An ultra-short-term and short-term wind power forecasting approach based on optimized artificial neural network with time series reconstruction. In: 2022 4th International Conference on Smart Power and Internet Energy Systems (SPIES). Beijing, China: IEEE; 2022. pp. 2068-2073. DOI: 10.1109/SPIES55999.2022.10082352

[88] Ponkumar G, Jayaprakash S, Kanagarathinam K. Advanced machine learning techniques for accurate ver-short-term wind power forecasting in wind energy systems using historical data analysis. Energies. 2023;**16**:5459. DOI: 10.3390/en16145449

[89] Faruque MO, Hossain MA, Islam MR, Alam SMM, Karmaker AK. Very short-term wind power forecasting for real-time operation using hybrid deep learning model with optimization algorithm. Cleaner Energy Systems. 2024;**9**:100129. DOI: 10.1016/j.cles.2024.100129

[90] Hossain MA, Chakrabortty RK, Elsawah S, Ryan MJ. Very short-term forecasting of wind power generation using hybrid deep learning model. Journal of Cleaner Production. 2021;**296**:126564. DOI: 10.1016/j.clepro.2021.126564

[91] Shang L, Miao L, Shan Y, Gnyawali KR, Zhang J, Kattel G. A hybrid ultra-short-term and short-term wind speed forecasting method based on CEEMDAN and GA-BPNN. Weather and Forecasting. 2022;**37**:415-428. DOI: 10.1175/WAF-D-21-0047.1

[92] Bousla M, Belfkir M, Haddi A, El Mourabit Y, Bossoufi B. Comparison of artificial intelligence approaches for estimating wind energy production: A real-world case study. Results in Engineering. 2024;**24**:103626. DOI: 10.1016/j.rineng.2024.103626

[93] Chen Y, Hu X, Zhang L. A review of ultra-short-term forecasting of wind power based on data decomposition-forecasting technology combination model. Energy Reports. 2022;**8**:14200-14219. DOI: 10.1016/j.egyr.2022.10.343

[94] Zhu A, Zhao Q, Wang X, Zhou L. Ultra-short-term wind power combined prediction based on complementary ensemble empirical mode decomposition, whale optimization algorithm, and Elman network. Energies. 2022;**15**(9):3055. DOI: 10.3390/en15093055

[95] He X, Lei Z, Jing H, Zhong R. Short-term probabilistic forecasting method for wind speed combining long-short-term memory and Gaussian mixture model. Atmosphere. 2023;**14**(4):1-16. DOI: 10.3390/atmos14040717

[96] Deng Z, Zhang X, Li Z, Yang J, Lv X, Wu Q, et al. Probabilistic prediction of wind power based on improved Bayesian neural network. Frontiers in Energy Research. 2024;**11**:1-11. DOI: 10.3389/fenrg.2023.1309778

[97] Yassen MA, El-Kenawy ESM, Abdel-Fattah MG, Ismail I, Mostafa HEDS. Explainable artificial intelligence for wind power forecasting model based on long short-term memory. Neural Computing and Applications. 2025;**37**:14589-14611. DOI: 10.1007/s00521-025-11230-5

[98] Reja RK, Amin R, Tasneem Z, Abhi SH, Bhatti UZ, Sarker SK, et al. A

new ANN technique for short-term wind speed prediction based on SCADA system data in Turkey. Atmosphere. 2023;**14**:10. DOI: 10.3390/atmos14101516

[99] Dolara A, Grimaccia F, Leva S, Mussetta M. Weather-based machine learning technique for day-ahead wind power forecasting. In: 2017. IEEE 6th International Conference on Renewable Energy Research and Applications (ICRERA). San Diego, CA, USA: IEEE; 2017. pp. 206-209. DOI: 10.1109/ICRERA.2017.8191267

[100] Jamil J, Mansouri M, Mimouni MF. Short time forecasting for wind power generation using artificial neural network. In: 19th International Multi-Conference on Systems, Signals and Devices (SSD). Algeria: Setif; 2022. pp. 321-325. DOI: 10.1109/SSD54932.2022.9955856

[101] Ates KT. Estimation of short-term power of wind turbines using artificial neural network (ANN) and swarm intelligence. Sustainability. 2023;**15**: 13572. DOI: 10.3390/su151813572

[102] Kishore B, Kabilan K, Rahul LS, Prajwal Priyadarshan G, Vishnutheerth EP, Rahul S, Mohan LK. Advancing short-term wind power forecasting by AI-driven models for improved accuracy. Electrical Engineering. 2025; **107**(11):14795-14810. DOI: 10.1007/s00202-025-03295-1

[103] Liu X, Zhou J, Qian H. Short-term wind power forecasting by stacked recurrent neural networks with parametric sine activation function. Electric Power Systems Research. 2021; **192**:107011. DOI: 10.1016/j.epsr.2020.107011

[104] Cali U, Sharma V. Short-term wind power forecasting using long-short term memory based recurrent neural network model and variable selection. International Journal of Smart Grid and Clean Energy. 2019;**8**(2):103-110

[105] Kannan S, Prabakaran D, Kumar DS, Sivaram S. A deep learning-based convolution neural networks to forecast wind energy. In: International Conference on Recent Trends in Electronics and Communication (ICRTEC), Mysore, India. 2023. pp. 1-6. DOI: 10.1109/ICRTEC56977.2023.10111917

[106] Song K, Xiao X, Ma W, Liu K. Wind power prediction based on long short-term memory neural network. In: 2rd International Conference on Neural Networks, Information and Communication Engineering (NNICE), Guangzhou, China. 2023. pp. 37-42. DOI: 10.1109/NNICE58320.2023.10105768

[107] Dev U, Ahmad S, Uddin MR, Hazari MF, Kabiruzzaman M, MSR Z. Day ahead wind energy forecasting in two distinct locations of Bangladesh using deep learning models. In: International Conference on Electrical, Computer and Communication Engineering (ECCE). Chittagong, Bangladesh. 2025. pp. 1-6. DOI: 10.1109/ECCE64574.2025.11013432

[108] Sankari S, Paramasivan SK, Arockiasamy K, Senthivel S, Thangavel M. Deep learning for wind speed forecasting using Bi-LSTM with selected features. Intelligent Automation and Soft Computing. 2023;**35**(3): 3829-3844. DOI: 10.32604/iasc.2023.030480

[109] He X, Ma Y, Xie J, Zhang G, Xie T. Enhanced wind power forecasting using graph convolutional networks with ramp characterization and error correction. Energies. 2025;**18**(11):2763. DOI: 10.3390/en19112763

[110] He Y, Li H. Probability density forecasting of wind power using quantile regression neural network and kernel density estimation. Energy Conversion and Management. 2018;**164**:374-384. DOI: 10.1016/j.enconman.2018.03.010

[111] Li L, Yin XL, Jia XC, Sobhani B. Day ahead powerful probabilistic wind power forecast using combined intelligent structure and fuzzy clustering algorithm. Energy. 2020;**192**:116498. DOI: 10.1016/j.energy.2019.116498

[112] Ibargüengoytia-Gonzáles PH, Reyes Ballesteros A, Borunda M, García-López UA. Prediction of wind power generation with modern artificial intelligence technology. Journal of Autonomous Intelligence. 2021;**4**(2):52. DOI: 10.32629/jai.v4i2.501

[113] Xiao H, He X, Li C. Probability density forecasting of wind power based on transformer network with expectile regression and kernel density estimation. Electronics. 2023;**12**(5):1187. DOI: 10.3390/eelectronics12051187

[114] Qureshi S, Shaikh F, Kumar L, Ali F, Awais M, Gürel AE. Short-term forecasting of wind power generation using artificial intelligence. Environmental Challenges. 2023;**11**: 100722. DOI: 10.1016/j.envc.2023.100722

[115] Mao J, Zhao J, Zhang H, Gu B. A novel hybrid deep learning model for day-ahead wind power interval forecasting. Sustainability. 2025;**17**(7):3239. DOI: 10.3390/su12073239

[116] Yang F, Feng J, Wang B, Zhang B, Li Z. Short-term wind power prediction by hybrid neural network based on information decomposition. In: 36th Chinese Control and Decision Conference (CCDC), Xian, China. 2024. pp. 3789-3794. DOI: 10.1109/CCDC62350.2024.10587536

[117] Yesilbudak M, Benli M. Current status of medium-term and long-term wind power forecasting in the literature. In: 13th International Conference on Renewable Energy Research and Applications (ICRERA). Japan: Nagasaki; 2024. pp. 1615-1620. DOI: 10.1109/ICRERA62673.2024.10815122

[118] Das A, Reddu DP, Baidya Roy S. Improving the skill of subseasonal to seasonal (S2S) wind speed forecasts over India using statistical and machine learning methods. 2024. Journal of Geographysical Research: Machine Learning and Computation, 1, e2024JH000187. DOI:10.1029/2024JH000187

[119] Colak I, Sagiroglu S, Yseilbudak M. Data mining and wind power prediction: A literature review. Renewable Energy. 2012;**46**:241-247. DOI: 10.1016/j.renene.2012.02.015

[120] Wu GZ, Zhang YB, Su C, Liu YJ. Study on medium-term and short-term wind power forecasting methods. Applied Mechanics and Materials. 2013; **361-363**:318-322. DOI: 10.4028/www.scientific.net/AMM.261-363.318

[121] Sobolewski RA, Tchakorom M, Couturier R. Gradient boosting-based approach for short- and medium-term wind turbine output power prediction. Renewable Energy. 2023;**203**:142-160. DOI: 10.1016/j.renene.2022.12.040

[122] Ekinci G. Ozturk HK Forecasting wind farm production in the short, medium and long terms using various machine learning algorithms. Energies. 2025;**18**(5):1125. DOI: 10.3390/en18051125

[123] Borunda M, De la Cruz J, Garduno R, Nicholson AE. Technical

assessment of small-scale wind power for residential use in Mexico: A Bayesian intelligence approach. PLoS One. 2020; **15**(3):e0230122

[124] Ge C, Yan J, Song W, Zhang H, Wang H, Li Y, et al. Middle-term wind power forecasting method based on long-span NWP and microscale terrain fusion correction. Renewable Energy. 2025;**240**:122123. DOI: 10.1016/j.renene.2024.122123

[125] Doblas-Reyes FJ, García-Serrano J, Lienert F, Biescas AP, Rodrigues LRL, Predictability SC, et al. Climatic Change. 2013;**4**:245-268. DOI: 10.1002/wcc.217

[126] Lledó L, Torralba V, Soret A, Famon J, Doblas-Reyes F. Seasonal forecasts of wind power generation. Renewable Energy. 2019;**143**:91-100. DOI: 10.1016/j.renene.2019.04.135

[127] Mbugua JM, Hiraga Y. Recent advances in long-term wind-speed and -power forecasting: A review. Climate. 2025;**13**(8):155. DOI: 10.3390/cli13080155

[128] Shobana DA, Maragatham G, Boopathi K, Lavanya MC, Sarangy R. Long-term wind speed forecasting – A review. In: Artificial Intelligence Techniques for Advanced Computing Applications. Singapore: Springer; 2020. pp. 79-99. DOI: 10.1007/978-981-15-5329-5_9

[129] Wu YK, Hong JS. A literature review of wind forecasting technology in the world. In: Proceeding of the 2007 IEEE Lausanne Power Tech, Lausanne, Switzerland, 1–5 July 2007. 2007. pp. 504-509. DOI: 10.1109/PCT.2007.4538368

[130] Pereis AT, Jayasinghe JMJW, Rathnayake U. Forecasting wind power generation using artificial neural network: "Pawan Danawi" – A case study from Sri Lanka. Journal of Electrical and Computer Engineering. 2021:1-10. DOI: 10.1155/2021/5577547

[131] Azad HB, Mekhilef S, Ganapathy VG. Long-term wind speed forecasting and general pattern recognition using neural networks. IEEE Transactions on Sustainable Energy. 2014;**5**:546-553. DOI: 10.1109/TSTE.2014.2300150

[132] Wang X, Liu Y, Hou J, Wang S, Yao H. Medium-and-long term wind-power forecasts considering regional similarities. Atmosphere. 2023;**14**(3): 430. DOI: 10.3390/atmos14030430

[133] Petkovic D, Shamshirband S, Anuar NB, Saboohi H, Wahab AWA, Protic M, et al. An appraisal of wind speed distribution prediction by soft compupting methodologies: A comparative study. Energy Conversion and Management. 2014;**84**:133-139. DOI: 10.1016/j.enconman.2014.04.010

[134] Yaghoubirad M, Azizi N, Farajollahi M, Ahmadi A. Deep learning-based multistep ahead wind speed and power generation forecasting using direct method. Energy Conversion and Management. 2023;**281**:116760. DOI: 10.1016/j.enconman.2023.116760

[135] Borunda M, Ramírez A, Garduno R, García-Beltrán CD, Mijares R. Enhancing long-term wind power forecastig by using an intelligent statistical treatment for wind resource data. Energies. 2023; **16**(23):7915. DOI: 10.3390/en16237915

[136] Yassen MA, El-Kenawy ESM, Abdel-Fattah MG, et al. Explainable artificial intelligence for wind power forecasting model based on long short-term memory. 2025. Neural Computing and Applications. 2025;**37**:14589-14611. DOI: 10.1007/s00521-025-11230-5

Chapter 3

Vertical Axis Wind Turbines Intelligent Farms

Fred Nitzsche

Abstract

The need of making wind farms more efficient demands the assessments of new concepts, bringing these to the demonstration level. This chapter targets more specifically floating wind farms of Vertical Axis Wind Turbines (VAWTs). These wind farms are specifically suitable for harvesting wind energy at deep-sea waters, the next frontier of offshore wind farms. The chapter describes the new concept of making clusters of three VAWTs to increase the overall average power output of wind farms. The cluster topology concept takes advantage of the acceleration of the flow observed through the gap between closely spaced turbines ("funneling" effect). The concept also involves the adoption of active blade pitch control strategies to both generate wakes deflections, optimizing the "funneling" effect (collective and cyclic control), and improve structural fatigue life using the higher-harmonic control technique for mitigating blade high dynamic response due to the aeroelastic effects.

Keywords: wind farms, deep water, vertical axis wind turbines, clustering topology, active blade pitch control, power output improvement, vibration attenuation

1. Introduction

Horizontal Axis Wind Turbines (HAWTs) are today the most common turbines. Vertical Axis Wind Turbines (VAWTs), however, conditionally on the type of wind farm installation, may have significant advantages over HAWTs. VAWTs have similar power-conversion performance than HAWTs per blade swept area. However, a VAWT design is intrinsically lighter due to the positioning of the electric generator at the base of the turbine. With this, the necessity of construction of massive towers to hold the nacelle with the generator and gearbox at the top of the installation is avoided. Furthermore, the VAWT operation is independent of the wind direction, a characteristic that is only achieved in the HAWT with the yaw orientation mechanism that adds up to the weight sustained by the tower. VAWTs received, during the years following the first oil crisis in the 1970's, great attention from the industry. Extensive wind farms with this type of turbine were established, particularly in the Mohave Desert in the USA. In Canada, at Cap-Chat in the Gaspe Peninsula, the largest (Darrieus-type) VAWT, 4 MW in power, 110 m tall, was built in the early 1980's (**Figure 1**).

Figure 1.
Cap-Chat (Quebec, Canada) Darrieus vertical axis wind turbine.

However, the Darrieus design was slowly abandoned in the following decades due to the complications associated with high vibration of the blades—aeroelastic forced response, and, consequently, fatigue problems. Several turbines suffered catastrophic failures under high winds. Therefore, any solution contemplated by modern designs should address this issue as well.

Offshore wind turbine technology becomes mandatory with the need to increase turbine size as wind energy becomes economically more competitive when generated by large turbines. Although there is growing awareness of the population on the need for renewable energy conversion, there is also a clear resistance to locating large wind turbines onshore or near-shore. Offshore horizontal axis wind turbines, for example, in the North Sea, are being installed currently on foundations on the seabed as shallow waters so allows. However, it is predicted that there will be saturation of the producing areas in these shallow waters. In other parts of world, shallow waters do not exist at all, like in many coastal regions to the Pacific Ocean. When this happens, the need for floating deep-water facilities will become necessary (**Figure 2**).

The installation of floating HAWTs is challenging because in a good design the center of gravity should be set as close to the sea surface as possible to ensure the hydrodynamic stability of the floating unit. This design could be achieved with HAWTs keeping the turbine generator close to the water surface and introducing a torque transmission system inside the tower, from the generator to the nacelle. However, this system adds substantial weight to the unit. On the other hand, VAWTs,

Figure 2.
Shallow vs. deep waters wind turbine installations.

exactly for naturally allowing the positioning of the electric generator at their base, will then become one viable alternative to HAWTs. In fact, VAWT Darrieus-type rotors received special attention for deep-water offshore floating applications in the 7th European Commission Framework Program with the DeepWind project as in Ref. [1] (**Figure 3**).

2. Wind turbine intelligent farms: Clustering topology concept

Contrarily to the general perception that wakes generated by upstream turbines always interfere negatively on the airflow reaching downstream wind turbines, hence reducing the overall energy production of wind farms unless a considerable distance between farm rows is verified, a recent investigation on closely spaced wind turbines, supported by both numerical calculations and wind tunnel tests, demonstrated that an increase in the energy output can be actually achieved as in Ref. [2]. This investigation was realized using HAWTs but the principle can be extended to VAWTs. For this, the turbines wake shedding needs to be positively explored by clustering. Clustering allows not only a higher density of turbines but also an average increase in the energy production output if the right topologic conditions are met. The physical explanation for the phenomenon is the "funneling" effect that the wakes of two closely spaced wind turbines located upstream create on the flow, accelerating it downstream (**Figure 4**).

The increase is verified to be very significant at certain dimensionless spacing distances between turbines, X/D and Y/D as defined in **Figure 4** (where D is the rotor diameter).

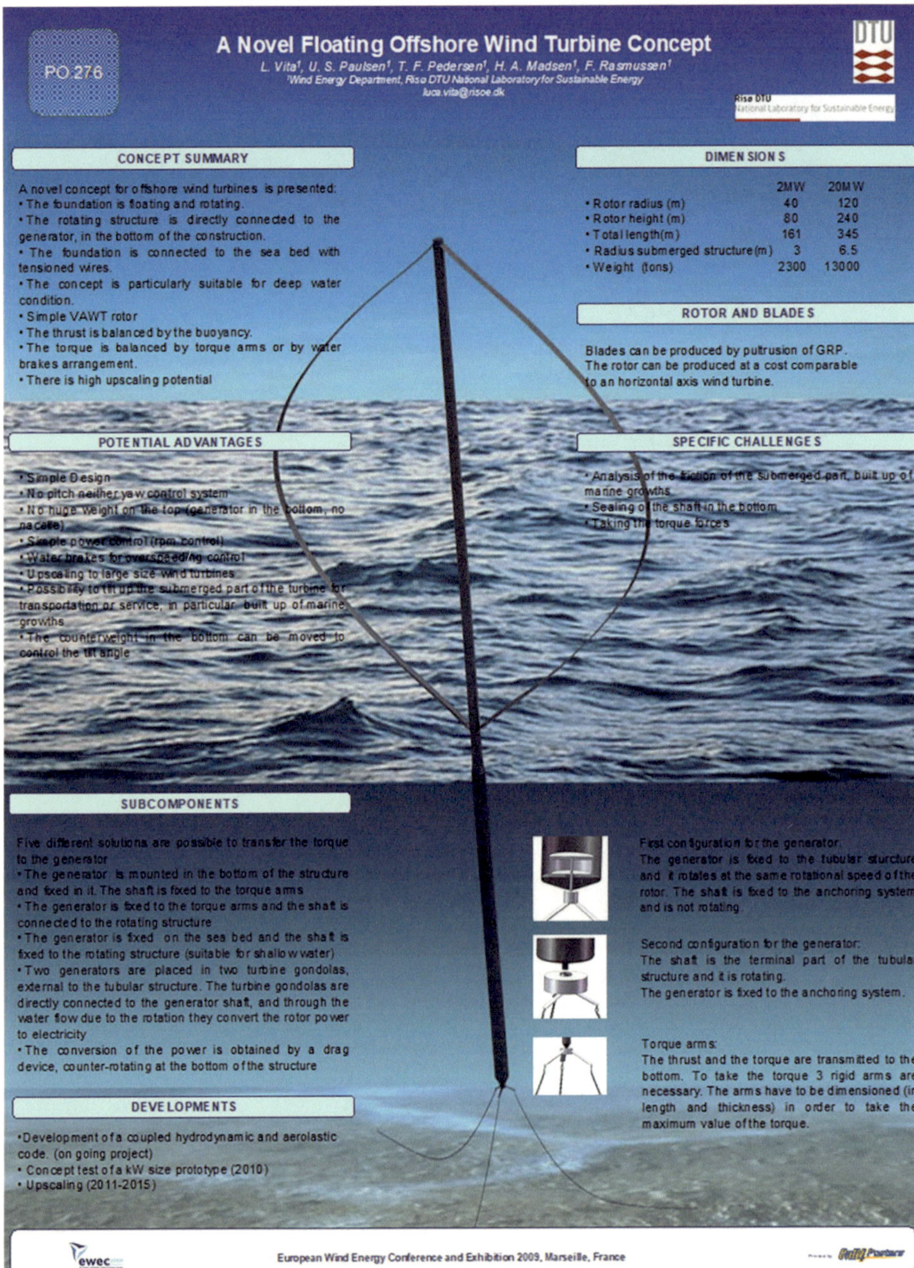

Figure 3.
DeepWind project (7th European Framework Program) floating Darrieus rotor concept. Poster from Danish Technical University (DTU) presented at European Wind Energy Conference and Exhibition, 2009, Marseille, France; reproduced with permission.

In particular, according to references [2, 3], **Figure 5** indicates that for the cluster configuration:

- Turbine 2 exhibits an increase in power output when located 1.0D to 2.5D downstream of the upstream rotors and exhibits a maximum increase in power output of 9% with a 1.0D longitudinal setback and a 3.0D gap;

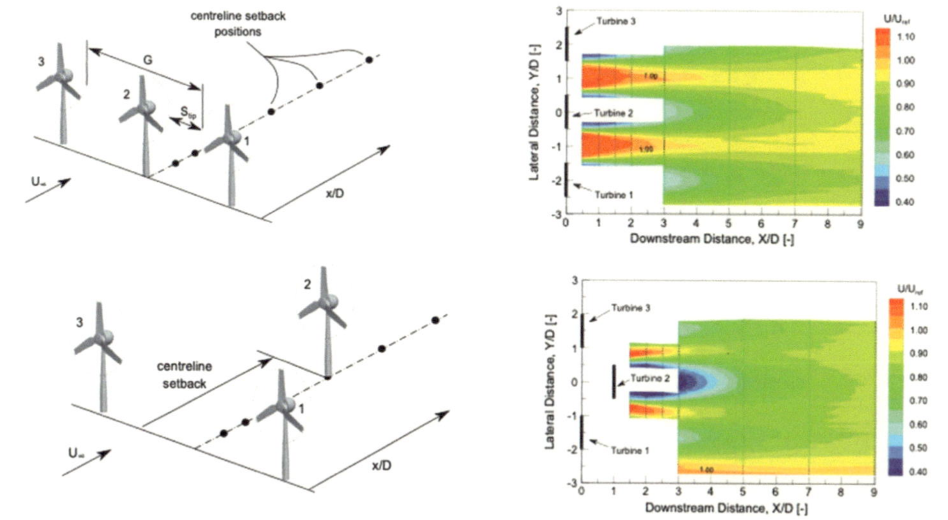

Figure 4.
Wake flow "funneling" effect observed in clusters of three closely spaced wind turbines: the three turbines in a single row configuration (upper left); flow acceleration is observed across the blades gaps of width Stip *(upper right). Same effect when the three turbines are placed along two rows (lower left). The two upstream turbines accelerate the flow toward the downstream turbine through the gap (lower right). From reference [3].*

- Turbine 2 experiences a maximum increase in power output of 10.5% with a 1.5D longitudinal setback, but this benefit is reduced as the rotor is moved further downstream. Turbines 1 and 3 also exhibit a notable increase in power output when the setback of the center turbine is minimized;

- Turbine 2 produces a maximum power increase of 7% with a 1.0D longitudinal setback, but exhibits a smaller overall increase in power compared to the previous case.

Interestingly, turbines 1 and 3 also exhibit a notable increase in power output when the setback of the center turbine (Turbine 2) is decreased. Therefore, the power net output of the cluster *of the three turbines* is likewise increased well over 10% when this particular cluster topology is established. Although cases (a) and (c) in **Figure 5** also indicate the substantial improvements for different cluster topologies, the result is highly dependent on the undisturbed wind speed, U_∞.

For different wind speeds a different spacing S_{tip} between turbines would be necessary to produce the similar improvements in the overall energy production of the cluster. However, the strength of the vortices generated by the wind turbine blades and, consequently, the wakes produced by the wind turbine itself are regulated by the angle of the attack of the blades (or blade pitch) for a given wind speed. Therefore, the aforementioned improvement in the output power that was obtained for fixed blade pitch could be extended for different wind speeds, independently of the spacing between turbines, if the blade pitch could be actively regulated and, thereof, the turbine rotational speed and loading. Although blade pitch control is relatively new concept to wind turbines, it is a mature technology in the field of helicopter rotor design. In wind turbines, it has been proposed as way of reducing aerodynamic loads in very large HAWTs where the ground effect makes the undisturbed wind speeds very different at the upper and lower blade positions over a blade revolution in Ref. [4].

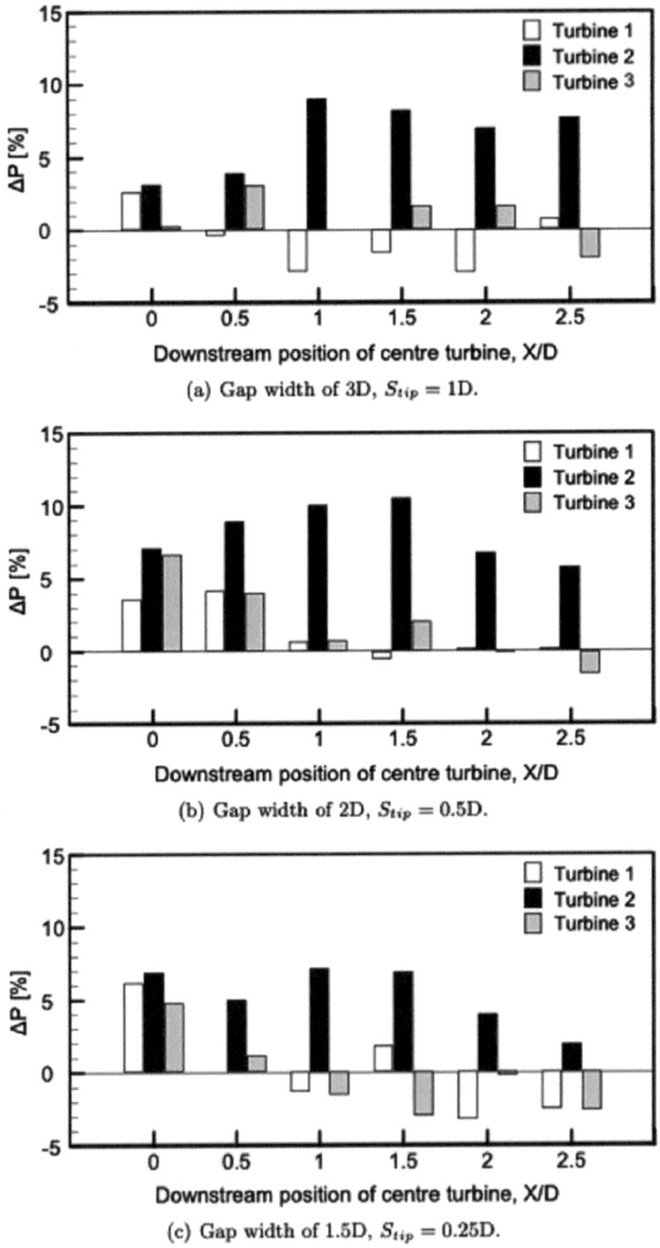

Figure 5.
Parametric study on the cluster centerline setback (ΔP is the power output increase for each turbine).

3. VAWTs blade pitch control: An analogy with helicopters in forward flight condition

In helicopters, the pilot controls the main rotor blades through the swashplate, which is a device that translates and tilts along the main rotor shaft following the hydraulic commands from the cockpit. **Figure 6** shows a schematic of the swashplate

DOI: http://dx.doi.org/10.5772/intechopen.1013118

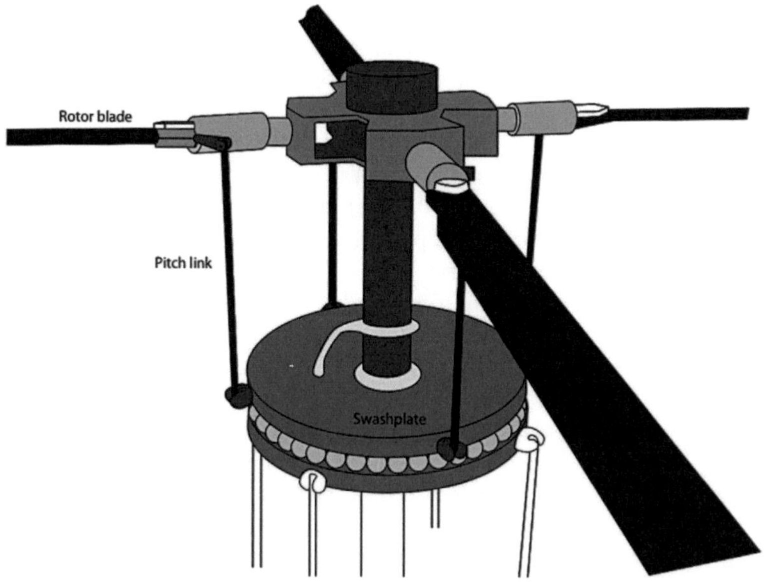

Figure 6.
The helicopter collective and cyclic flight controls with the swashplate device.

with its pitch link attachments to the rotor blades. The pilot controls hydraulic actuators that move the lower, non-rotating plate, while the upper plate rotates with the rotor head-on bearings housed between them. The pilot can both command these plates vertically and tilt them with the controls in the cockpit. The vertical motion of the swashplate induces a *collective pitch* at each blade, whereas its tilt orientation determines the corresponding *cyclic pitch*. Theoretically, a pair of plates composes the swashplate: one in the rotating frame and another in the helicopter body-fixed frame. Both are centered on the main rotor shaft and free to translate vertically along the shaft as a single unit. The vertical position and orientation of the swashplate determine the total pitch of the rotor blades, which has then two components superimposed: (1) the baseline pitch of the rotor blades called collective, or collective pitch. The collective pitch controls all rotor blades angles of attack equally, independently of their azimuth positions. (2) The cyclic pitch, governed by the tilt of the swashplate with respect to the rotor shaft, commands the angle of attack of each rotor blade differently but in a cyclic pattern that varies according to its azimuthal position. Hence, the collective pitch is a rigid-body motion in the rotating frame performed at a normalized frequency 0/rev, whereas the cyclic pitch is done at a normalized frequency 1/rev, where "rev" intends for the rotor revolution frequency. Blade pitch control can be also done at frequencies higher than 1/rev. In helicopter design, this type of control is called Higher-harmonic Control or for short HHC, as in Ref. [5]. It is intended to superimpose to the collective and cyclic blade pitch control frequencies higher harmonics p/rev. These harmonics are aimed to attenuate high-frequency signals present in the blades having dynamic responses to unsteady aerodynamic loads (aeroelastic forced response). They are especially significant in the forward-flight regime of helicopters, when there is strong interaction between the vortices shed by each blade individually on the remaining blades during one revolution, in a phenomenon called Blade-Vortex Interaction (BVI). These vibrations can not only create structural fatigue to blades and other components of the helicopter but also cause human discomfort when riding helicopters. The vibrations are transmitted to the fuselage through the rotor shaft pronouncedly at integer multiples of the rotor number of blades (N/rev), where $N = nB$ is the product of the rotor number of blades, B and n are integers, in Ref. [6]. The HHC concept (where all blades are controlled in unison by the same signal) can be generalized to the Independent Blade Control (IBC) where each blade is individually fed with independent signals, as in Ref. [7]. The control signals are produced either in open loop or in closed-loop automatic control utilizing sensors to continuously monitor the structural response.

Although to the knowledge of the author it has not been reported in the open literature, blade pitch control can also be introduced for VAWTs. In **Figure 7**, the basic aerodynamics of wind turbines is introduced. $\mathbf{U}_\infty$ is the free (unperturbed) wind speed blowing in the direction of the unit vector $\hat{\mathbf{e}}_1$. The turbine of radius, r_T has an angular velocity and Ω_T in the counterclockwise direction. The blade airfoil aerodynamics is indicated in the figure at four azimuth angles, θ_T referred to the wind speed direction, $\hat{\mathbf{e}}_1$ where $\theta_T = 0$. At each azimuth position, the airfoil tangential velocity is $\mathbf{V}$. Therefore, the airfoil composed vector velocity, $\mathbf{V} + \mathbf{U}_\infty$ produces the local instantaneous angle of attack, α_{geom}, the lift, $\mathbf{L}$ and drag, and $\mathbf{D}$ forces as indicated in the figure. These can be resolved onto the non-rotating frame: the lateral force, $\mathbf{F} = F_1\,\hat{\mathbf{e}}_1 + F_2\,\hat{\mathbf{e}}_2$ and the torque about the rotational axis,

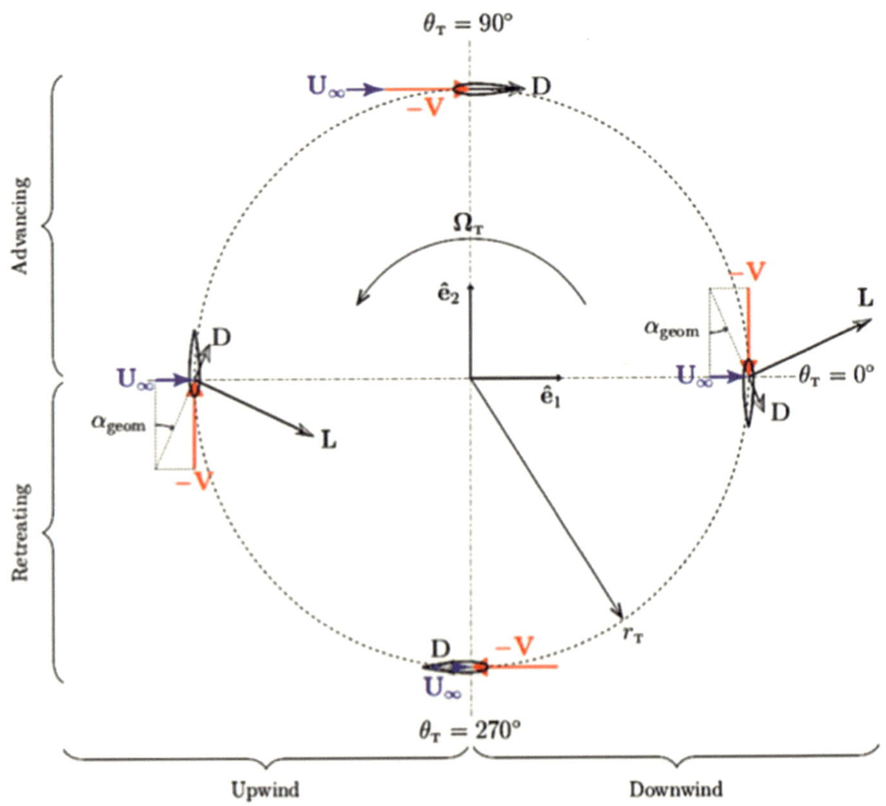

Figure 7.
VAWT basic bade aerodynamics. Situations in the upwind, downwind, advancing, and retreating quadrants of operation.

$\tau = \tau\,\hat{\mathbf{e}}_3$ (where the unit vector $\hat{\mathbf{e}}_3$ points upward the plane defined by $\hat{\mathbf{e}}_1$ and $\hat{\mathbf{e}}_2$). For VAWTs in normal operation, as long as Ω_T achieves a minimum value, the integral $\int_0^{2\pi} \tau\, d\theta_T > 0$. If blade pitch control is introduced using for example the HHC strategy, $\alpha_{geom}(\theta_T)\big]_{HHC} = \alpha_{geom}(\theta_T) + \sum_p \Delta\alpha_p(\theta_T)\cos pt$, where p can assume any integer value, namely 0 for collective, 1 for cyclic, and $p \geq 2$ for the higher-harmonic control frequency p/rev. Some harmonics may be skipped in HHC if control at that harmonic is not relevant for the dynamic response, for $p \neq nB$, for example. The result of HHC (or more generically IBC) would be that not only the torque could be actively controlled for a certain wind speed but also the lateral force. More interesting for the present development is that calling the flow conservation of momentum principle if the lateral force can be controlled likewise the direction of the turbine shed wake can. This fact can be used to explore the observed acceleration of the flow in the gaps between turbines operating in close proximity using the "funneling" effect referred in Section 2.

Two possible concepts for realizations of the blade pitch control are next presented for H-type, 2-blade rotors.

- Concept 1: As seen in **Figure 8**, it is straightforward to design a mechanism with equivalent functions of the swashplate for operation in H-type vertical axis wind turbines. Note that tilt of the plate is not need as the plane of rotation in VAWTs is already the one of blade actuation. The mechanism would be located midway the blades span at their horizontal connections with the vertical axis. Two rigid links would push-pull the blades off their pitch axes to provide the independent control of the angles of attack. In the sketch presented, two sets of linear electromechanical actuators are connected to the pitch links to send the independent signals for each blade (IBC concept). The pitch links are designed to amplify the motion supplied by the actuators. The control signals would be a superposition of the 0/rev (collective), 1/rev (cyclic), and higher-harmonic signals as well (2/rev and up) to perform blade load alleviation.

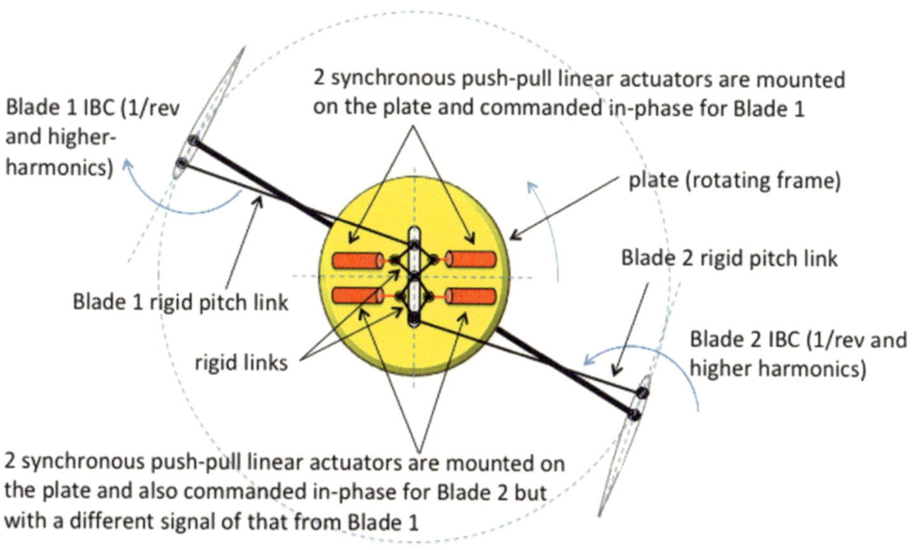

Figure 8.
Blade pitch control concept for a 2-blade H-type VAWT.

- Concept 2: A second and more innovating concept is depicted in **Figure 9**. As seen, it does not require linear actuators or other mechanisms to provide blade pitch independent control. Instead, it introduces a second generator in the turbine; one dedicated to each blade. Any electric generator has some swashplate features because, as the swashplate, it has non-rotating (the stator) and rotating frame (the rotor) components. As indicated, each generator rotor is attached to a separate tube that is connected the pitch link as in Concept 1. The two rotating tubes (one for each blade) run concentrically to the turbine shaft. Then, an electronic device would control a constant plus a time-dependent phase difference between the two generator rotors to supply the required collective and cyclic pitch controls of the two blades and the high-harmonic IBC signal during the revolution.

The same concept could also be extended to floating Darrieus VAWTs introduced in **Figure 3**. The actuation scheme is depicted in **Figure 10**. In the Darrieus, compared to other designs, the blades (of length L seen the figure) are very flexible. This is because their shape approximates the "troposkien" curve, which is the shape of a rotating rope. With this, the blades transversal stresses are minimized, as noted in Ref.s [8, 9]. Differently to the H-type turbine, this type of rotor blade has two boundary conditions. The proposed pitch control would, in this case, be performed at the both ends of the blade, as indicated in the figure, maintaining the characteristic lightweight design of the rotor. As a result, the two blades would be simultaneously twisted (morphed) by the same angle applied at their roots but in opposite directions, $-\Delta\alpha(0,t)$ and $+\Delta\alpha(L,t)$ to provide the local, time-dependent angle of attach for IBC, $\Delta\alpha(x_1,t)$.

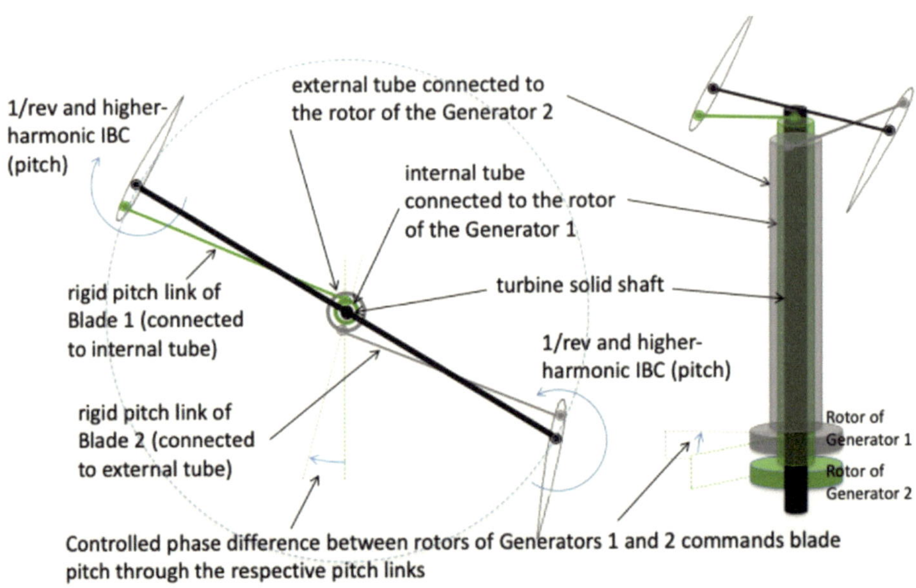

Figure 9.
Blade pitch control alternative concept for a 2-blade H-type VAWT. Actuation using two generators dedicated to each blade individually.

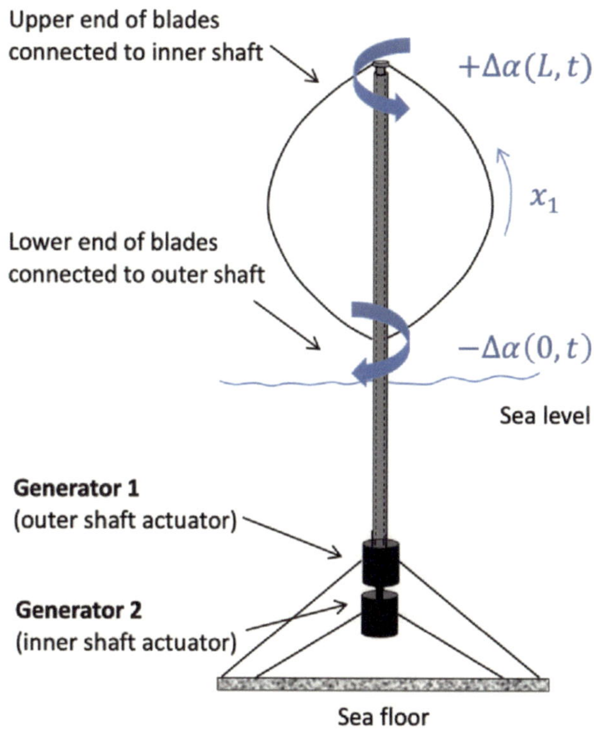

Figure 10.
Floating Darrieus rotor with blade pitch control.

4. Intelligent clustering topology extended to VAWTs: Challenges

The main purpose here is to present, as seen by this author, the current challenges to design intelligent wind farms of VAWTs using the clustering topology described in Section 2. For this, the extension of the previous cluster results obtained for HAWTs to VAWTs using reduced-scale models, for example, to minimize experimental costs, should be first demonstrated. Second, as the energy output of the cluster is fundamentally dependent on the characteristics of the wake shed by the upstream turbines, incorporation the blade pitch control, as discussed in Section 3, is necessary. This strategy will be able to tune both the wake strength at an instantaneous wind speed and its direction using the collective and cyclic controls. The active control system would be able to maximize the cluster energy output within broad range of wind speeds, which is the main objective of an intelligent wind farm. As mentioned in Section 1, high level of blade vibration, especially in the Darrieus rotor case, was the plague that prevented their industrial dissemination in the past as reported in Ref. [10]. It became then obvious that some type of active blade control should be introduced to realize more robust designs. A previous work from one of the author's students, using the vortex-particle unsteady aerodynamics method to represent the free-wake shedding, and the finite element method for the flexible structural representation, aeroelastic simulations of a single 2-blade Darrieus rotor were performed in Ref. [11] (**Figure 11**). The result clearly confirms that the structural response of the blade is dominated by the multiples of the number of blades, with 2/rev the highest component in the spectrum, as seen in **Figure 12**. In the proposed cluster

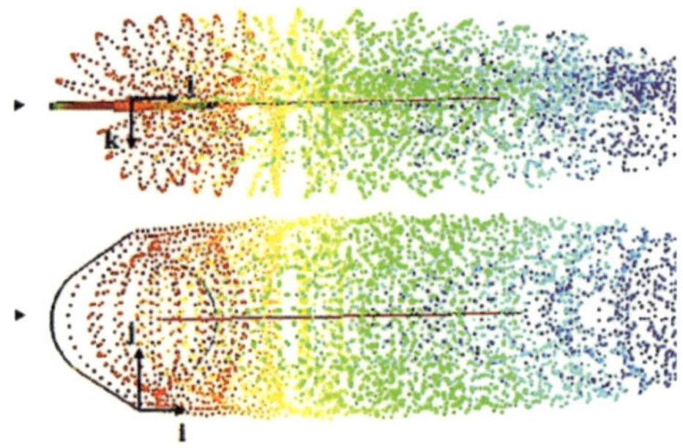

Figure 11.
Aeroelastic simulation of a 2-blade Darrieus rotor showing the free-wake shedding (wake colors indicate local strength).

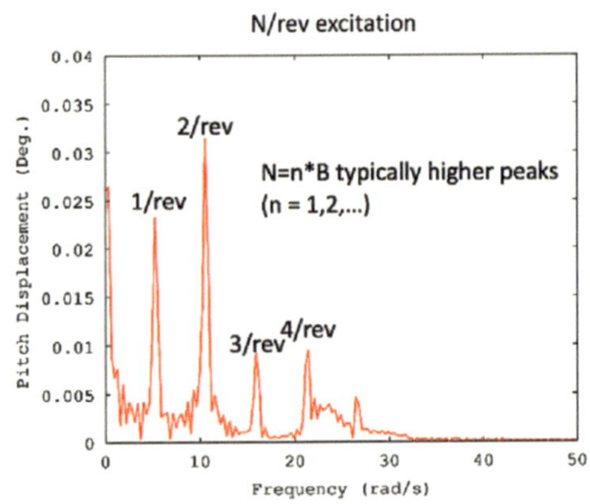

Figure 12.
Darrieus structural blade response spectra (torsional degree of freedom).

configuration, to optimize the energy production, turbine proximity is a requirement. Therefore, especially at high wind speeds, active HHC or IBC systems presented in Section 3 would bring clear benefits into the more modern designs.

The cluster topology taking advantage of the "funelling" effect discussed in Section 2 is extended for VAWTs in **Figure 13**. It is worthwhile to note that the rotation direction of the upstream turbines does not need to be reversed to produce desirable effects on the downstream turbine. The change in the direction of wake shedding is achieved by changing the blade pitch cyclic control schedule over one turbine revolution. As a result of the cyclic control, the magnitude of lateral forces generated over one cycle of revolution and their direction can be changed as well for a fixed gap distance, G. Conservation of momentum of the fluid, thus, guarantees the deflection of the wakes as suggested in **Figure 13**.

This follows in close analogy to what happens in helicopters, which can fly in all directions (i.e., change the direction of the thrust vector) by adjusting their cyclic control schedule. Therefore, the presented strategy can provide the right response of the cluster to changes in the free wind conditions; at "lower" wind speeds, when the deflection of the wakes of the upstream turbines toward the downstream turbine is expected to produce the beneficial effects, that is, to increase in the power output taking advantage of the "funneling" effect without significantly affecting the turbine vibration levels, the wake direction is kept as indicated in **Figure 13**. However, for "higher" wind speeds a better

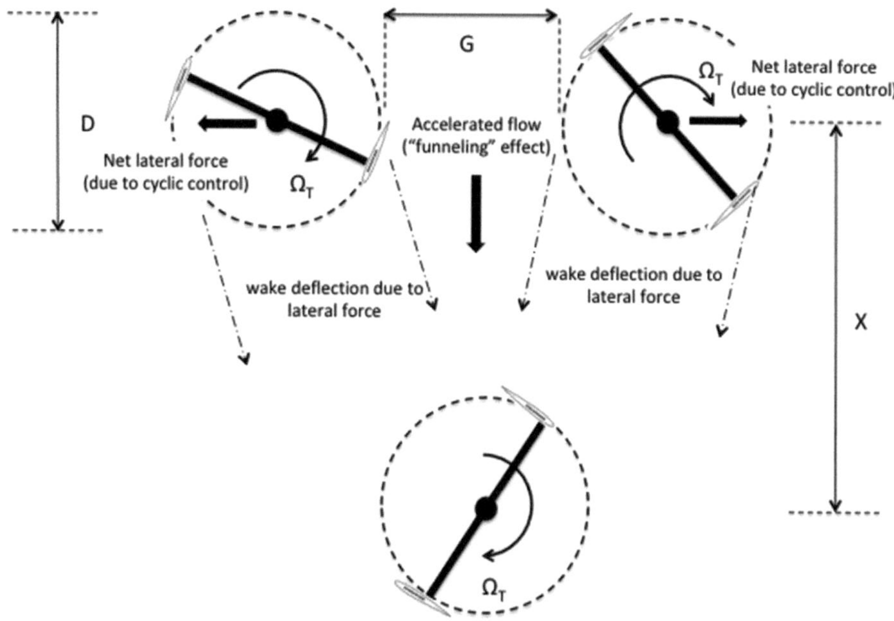

Figure 13.
VAWT cluster concept for maximizing the energy production using the funneling flow effect.

strategy would be to deflect the wakes produced by upstream turbines outward and away from the downstream turbine to reduce vibration.

Finally, a hierarchical control concept that at its *low-level* mitigates vibrations using higher-harmonic blade pitch control and at its *high-level* optimizes the cluster energy output is presented. The conceptual scheme of such a *multi-level* control system pursued by an intelligent VAWT wind farm is depicted in the block diagram on **Figure 14** and then further detailed on **Figure 15**.

- A central controller (high-level control) commands the blade pitch using the wind speed conditions measured with a Light Detection and Ranging (LIDAR) system [12] and determines the suitable control strategy, either for "low" or "high" winds. For this, (1) it receives fed back information from local turbine controllers installed on the turbines about the lateral force as a measure of the wake deflections and (2) the instantaneous cluster energy production. Then, it sets the blades collective and cyclic angles to maximize energy output based on a previously inputted Reduced-Order Model (ROM) obtained from offline flow simulations of the wind farm.

- A local controller (low-level control) receives input information from: (1) the central computer and (2) local sensors on the turbine torque, blades, and shaft loads and determines the higher-harmonic pitch angles for vibration alleviation.

Figure 15 further indicates that for providing energy to the local sensors that measure the torque, blade vibration, and shaft loads, energy harvesting from the blade vibrations themselves can be employed, making the turbine power generation output independent from the automatic control loop power requirements, that is, making it an autonomous control system. Nowadays, this is a mature technology, as seen for example in Ref. [13].

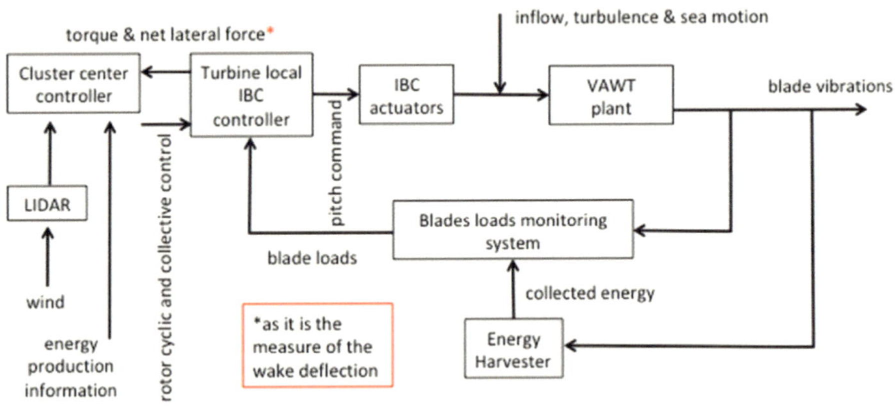

Figure 14.
Block-diagram of an intelligent VAWT wind farm.

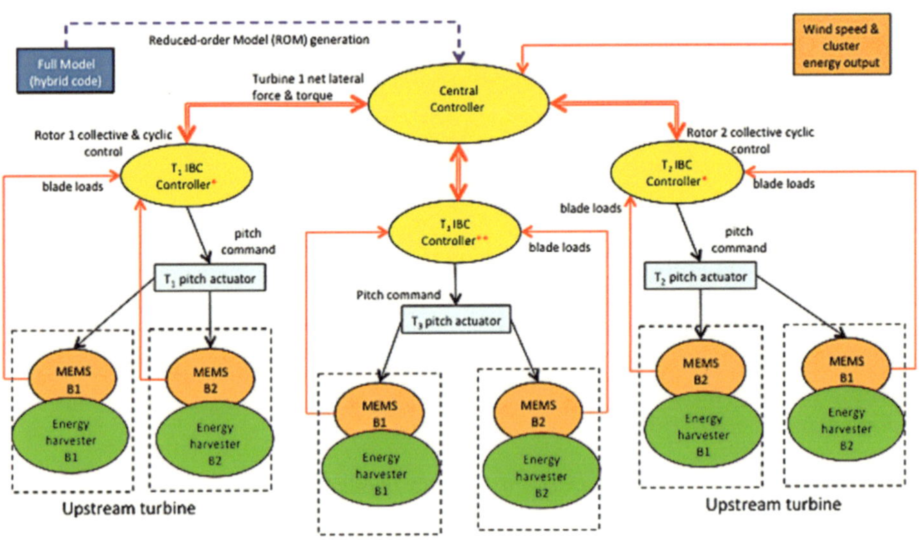

Figure 15.
Hierarchical multi-level control of an intelligent VAWT wind farm.

5. Conclusions

In the present chapter, a novel topological concept for developing floating wind farms of vertical-axis wind turbines and its challenges was presented. It is based on dividing the whole wind farm into clusters of three turbines where the units are closely spaced. It takes advantage of an observed phenomenon verified in both flow simulations and wind tunnel tests with reduced-scale models of horizontal-axis wind turbines farms. In this case, the close proximity of two units located upstream accelerates the flow throughout the gap between the turbines, improving the performance of a third downstream wind turbine that is centered in the middle of the gap. As the reported improvement was demonstrated to be dependent on the free wind speed, blade automatic pitch control (collective and cyclic) was introduced to make the phenomenon independent of the flow conditions. In addition, higher-harmonic control of the blade pitch is proposed to attenuate the turbine vibration, a problem verified in former vertical-axis wind turbine designs, which eventually led to the catastrophic failures by premature fatigue of the structural elements and, ultimately, prevented widespread use of this type of turbine. Possible realizations of the control mechanism, based on usual helicopter technology adapted to vertical-axis wind turbines, were suggested. Finally, a hierarchical, multi-level, control scheme for an intelligent wind farm exploring the novel concept was presented.

Acknowledgements

The author would like to thank throughout the 25 years of his tenure at Carleton University to his more than 50 graduate students, his research partners, and organizations particularly in Germany, Italy, Greece, and Denmark to valuable team contributions. The author would also like to thank the Canadian research funding organizations NSERC (National Science and Engineering Research Council) and OCE (Ontario Centers of Excellence) for support to different pieces of the research material presented.

Conflict of interest

The author declares no conflict of interest.

Nomenclature

B	number of rotor blades
$\mathbf{D}$	airfoil drag
D	rotor diameter
$\hat{\mathbf{e}}_i$	unit vector (ground-fixed frame) in the directions 1, 2, 3 (same as **i, j, k**)
$\mathbf{F}$	lateral force acting on the rotor
G	lateral gap between the tips of adjacent turbines located upstream
$\mathbf{L}$	airfoil lift,
L	blade length in the Darrieus rotor
n	integer number
N	particular harmonic of the rotor rotation frequency, $N = nB$
p	normalized higher harmonic control frequency
P	turbine power output
r_T	turbine radius
S_{tip}	lateral gap between the tips of the blades of adjacent turbines
t	time
$\mathbf{U}_\infty$	undisturbed wind velocity (= $\mathbf{U}_{ref}$)
$\mathbf{V}$	turbine tangential velocity
x_1	curvilinear spanwise coordinate in the Darrieus rotor blade
X	centerline setback distance between upstream turbines and downstream turbine
Y	longitudinal direction in the wind farm
/rev	normalized frequency (with respect to turbine rotational frequency)

α_{geom}	airfoil geometric angle of attack
$\Delta\alpha$	airfoil (blade) controlled angle of attack
θ_T	blade azimuth angle
Ω_T	turbine rotational angular velocity
τ	turbine torque

Acronyms

HAWT	horizontal-axis wind turbine
HHC	blade high-harmonic control
IBC	individual blade control
VAWT	vertical-axis wind turbine

Author details

Fred Nitzsche
Carleton University, Ottawa, Canada

*Address all correspondence to: fred.nitzsche@carleton.ca

References

[1] Paulsen US et al. Outcomes of the DeepWind conceptual design. In: Energy Procedia 12th Deep sea Offshore Wind R&D Conference, Trondheim, Norway; 4-6 February 2015. Amsterdam, NL: Elsevier Procedia; 2016

[2] McTavish S, Feszty D, Nitzsche F. A study of the performance benefits of closely-spaced lateral wind farm configurations. Renewable Energy. 2013;**59**:128-135

[3] McTavish S. Identification of wind turbine testing practices and investigation on the performance benefits of closely-spaced lateral wind farm configurations, (PhD thesis). Ottawa, Canada: Carleton University; 2013

[4] Bottasso CL, Croce A, Riboldi CED, Salvetti M. Cyclic pitch control for the reduction of ultimate loads on wind turbines. Journal of Physics: Conference Series. 2014;**524**. DOI: 10.1088/1742-6596/524/1/012063

[5] Patt D et al. Higher-harmonic control algorithm for helicopter vibration reduction revised. Journal of Guidance, Control, and Dynamics. 2005;**28**(5):918-930

[6] Johnson W. Helicoper Theory. NY, USA: Dover Publications; 1980

[7] Ham ND. Helicopter individual blade control at MIT 1977-1985. Vertica. 1987;**11**(1/2). 109:122

[8] Blackwell BF, Reis GE. Blade Shape for a Troposkien Type of Vertical-Axis Wind Turbine, Sandia National Laboratories Report SLA 74-0154, NM, April 1974. Albuquerque, NM, USA: Sandia National Laboratories; 1974

[9] Reis GE, Blackwell, BF. Practical Approximations to a Troposkien by Straight-Line and Circular-Arc Segments, Sandia National Laboratories Report SAND 74-0100, NM, March 1975. Albuquerque, NM, USA: Sandia National Laboratories; 1974

[10] Nitzsche F. Dynamic aeroelastic stability analysis of vertical-axis wind turbines under constant wind velocity. AIAA Journal of Propulsion and Power. 1994;**10**(3):348-355

[11] Fereidooni A. Numerical study of aeroelastic behaviour of a troposkien shape vertical axis wind turbine, (MASc dissertation). Ottawa, Canada: Carleton University; 2013

[12] Simley E, Pao LY. LIDAR Wind Speed Measurements of Evolving Wind Fields. NREL Report SR-5000-55516. Golden CO, USA; Jul 2012. Available from: https://docs.nrel.gov/docs/fy12ost/55516.pdf

[13] Hossain I et al. MEMS-based energy harvesting devices for low-power applications – A review. Results in Engineering. 2023;**19**. DOI: 10.1016/j.rineng.2023.101264

Chapter 4

Utilizing Relative Wind Energy for Dynamic Charging in Electric and Hybrid Vehicles

Cristian Helera and Dan Alexandru Stoichescu

Abstract

This chapter explores an innovative approach to extending the range of electric vehicles (EVs) by utilizing relative wind energy generated during motion to charge vehicle batteries. Traditional electric cars face challenges with battery capacity and frequent charging needs, limiting their travel range. By incorporating a wind turbine mounted on the vehicle chassis, it is possible to harness the wind generated as the vehicle moves, known as relative wind, to produce additional electric power. This solution aims to increase efficiency without introducing significant drag or reducing vehicle performance. The chapter discusses the aerodynamic considerations, digital simulations, and experimental results that demonstrate the feasibility of this approach, offering insights into a sustainable method for enhancing electric cars range and reducing dependency on external charging sources. The study highlights that integrating relative wind energy is a viable and sustainable method to extend hybrid car or electric car range, decrease reliance on external charging, and promote the use of renewable energy in the automotive sector.

Keywords: relative wind, electric vehicle (EV), wind turbine, battery charging, renewable energy, aerodynamic efficiency, energy conversion, sustainable mobility, automotive innovation, wind energy utilization

1. Introduction

A new commitment to reduce carbon emissions and dependency on fossil fuels is leading to a shift in the automobile sector. Future automotive technologies like hybrid and electric cars (EVs) may be able to assist in overcoming these obstacles. However, although being sustainable, these cars have a lot of disadvantages, particularly with regard to range anxiety and battery capacity. Although battery technology continues to advance, consumer expectations for driving range and charging rates are still not being fulfilled. The situation has led to research into alternative battery-free sources of energy that can increase the range of hybrids and electric cars. These issues are expected to be fixed by renewable energy sources such as solar and wind [1].

1.1 Concept of relative wind energy

Relative wind energy is defined as the wind created as a vehicle moves through the air. It can be captured using a wind turbine mounted on the vehicle body. The ideology works on the concept that as the vehicle moves, it creates a flow of air that can be captured and converted into electrical energy. This method is intended to be used as an additional power source for the vehicle battery, thus increasing the vehicle's range.

1.1.1 Mechanism of energy capture

Using a power source, the wind turbine transforms the relative wind's motion into electrical energy. The turbine's design, the speed of the car, and its aerodynamic characteristics are some of the variables that affect how efficiently this energy conversion takes place.

- Turbine design: the turbine blades are intended to catch the maximum amount of wind energy. The form, angle, and material of the blades all have an impact on capturing energy efficiency.

- Generator: the generator converts the mechanical energy from the turbine into electrical energy. The type of generator (e.g., permanent magnet synchronous generator) and its efficiency play a significant role in the overall system performance.

- Power electronics: power electronics are used to manage the electrical energy generated by the turbine. This includes rectifiers, inverters, and controllers that ensure the energy is efficiently stored in the vehicle's battery [2].

1.1.2 Benefits of utilizing relative wind energy

- Extended range: by generating additional power during motion, the vehicle's range can be extended without relying solely on battery capacity.

- Reduced charging frequency: the supplementary power reduces the need for frequent external charging, enhancing the convenience for users.

- Sustainable energy source: utilizing wind energy contributes to the overall sustainability of the vehicle by reducing dependency on non-renewable energy sources [3].

1.2 Importance of aerodynamic considerations

Integrating a wind turbine on a vehicle requires a careful understanding of aerodynamic principles to ensure that the turbine will not affect the vehicle's performance. The turbine's design and position must be adapted to reduce drag and increase the generation of energy.

1.2.1 Design challenges

- Minimizing drag: the turbine must be designed to minimize additional drag, which can negatively impact the vehicle's efficiency and performance. This involves optimizing the shape and placement of the turbine blades to reduce air resistance.

- Optimal placement: the placement of the turbine on the vehicle is crucial. It should be positioned in a way that captures the maximum amount of wind energy without disrupting the vehicle's aerodynamics. Common placements include the roof or rear of the vehicle, where airflow is more stable.

- Structural integration: the turbine must be integrated into the vehicle's structure in a way that does not compromise safety or stability. This includes ensuring that the turbine is securely mounted and that its weight distribution does not affect the vehicle's handling [4].

1.2.2 Digital simulations

Before actual implementation, computer simulations are performed to simulate the vehicle's aerodynamic behavior with the wind turbine. These models assist in determining the turbine's ideal placement and design. These models take wind speed and direction, as well as the vehicle's physical characteristics, into consideration.

- Simulation tools: advanced computational fluid dynamics (CFD) tools are used to simulate the airflow around the vehicle and the turbine. These tools provide detailed insights into how the turbine affects the vehicle's aerodynamics.

- The simulation method is iterative, with the turbine's design and placement modified based on the results of each simulation phase. This helps in achieving an appropriate balance of energy capture and aerodynamic efficiency.

1.3 Objectives of the chapter

The purpose of this chapter is to present the idea of dynamic charging electric and hybrid cars using relative wind energy. In order to set up the comprehensive simulations and experimental validations discussed in the following sections, it will explore the background, importance, and potential benefits of this approach. Among the goals are:

- understanding the need for renewable energy in vehicles: highlighting the challenges faced by EVs and hybrids and the potential of renewable energy to address these issues;

- exploring the concept of relative wind energy: explaining how wind energy can be harnessed by vehicles in motion and the benefits of this approach;

- emphasizing aerodynamic considerations: discussing the importance of optimizing turbine design and placement to ensure minimal impact on vehicle performance; and

- setting the stage for further research: providing a foundation for the detailed simulations and experimental affirmations that will be covered in the following chapters.

2. Digital simulations and design optimization

2.1 Simulation tools and techniques

Digital simulations are an important step in the development and optimization of wind turbines for electric cars. These simulations help predict the aerodynamic behavior of the vehicle with the turbine installed, allowing for adjustments to be made before physical prototypes are produced.

2.1.1 Computational fluid dynamics (CFD)

CFD tools are used for simulating the airflow around the vehicle and turbine. These methods solve complex equations that describe airflow, providing accurate data regarding how the turbine affects the vehicle's aerodynamics.

- Navier-Stokes equations: the fundamental equations governing fluid flow, which are solved numerically in CFD simulations.

- Turbulence models: models such as k-ε (k-epsilon) and LES (Large eddy simulation) are used to accurately predict turbulent flows around the vehicle and turbine.

- Mesh generation: creating a computational grid (mesh) that divides the vehicle and surrounding air into small elements for numerical analysis. A finer mesh provides more precise results, but it demands a greater amount of processing power [5].

2.2 Results and analysis

The simulation results are examined in order to optimize the turbine's design and positioning. This includes evaluating pressure and velocity characteristics, as well as performing structural studies to ensure the turbine's integrity.

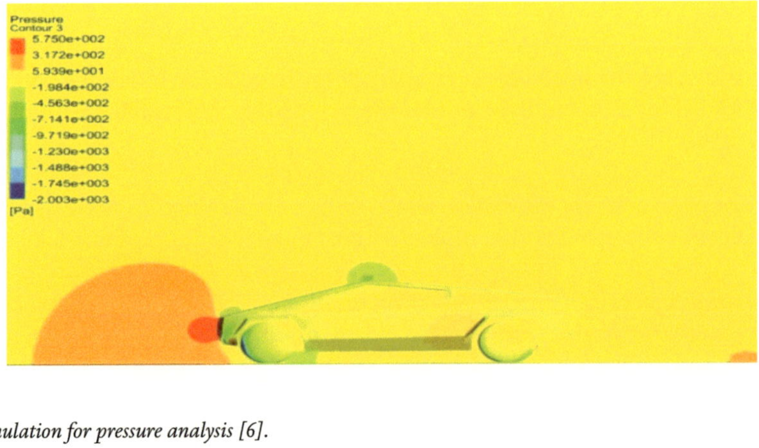

Figure 1.
Digital simulation for pressure analysis [6].

2.2.1 Pressure contours

Figure 1 illustrates the pressure contours around the vehicle and turbine, highlighting areas of high and low pressure. This information is essential for optimizing turbine positioning to minimize drag and optimize energy capture.

- High-pressure zones: areas where the airflow slows down and pressure increases, typically found at the front of the vehicle.

- Low-pressure zones: Areas where the airflow speeds up and pressure decreases, often found at the rear of the vehicle.

2.2.2 Velocity contours

Figure 2 shows the velocity contours around the vehicle and turbine, indicating regions of high and low wind speed. These contours help explain how the turbine interacts with the airflow and how much energy it can accumulate.

- High-velocity regions: Areas where the airflow is fast, indicating potential for high energy capture.

- Low-velocity regions: Areas where the airflow is slow, indicating less potential for energy capture.

2.2.3 Structural analysis

Structural research ensures that the turbine and mounting system are capable of absorbing the forces that will be generated during running. This includes both aerodynamic forces.

- Stress distribution: **Figure 3** provides a detailed view of how stresses are distributed across the car and mounting system.

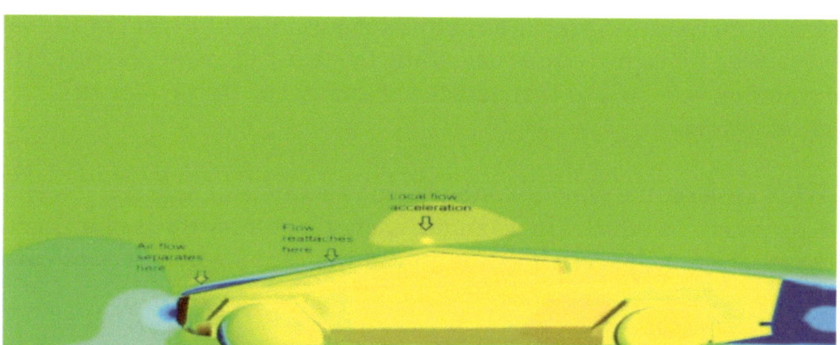

Figure 2.
Digital simulation for velocity analysis [6].

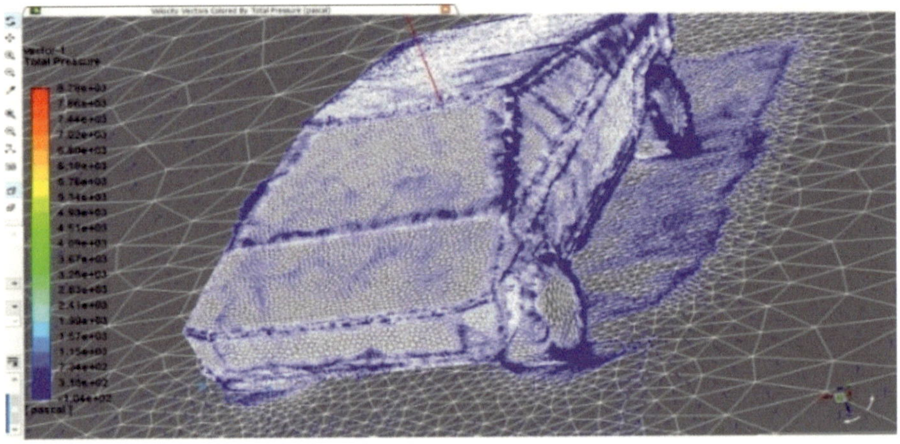

Figure 3.
The relative wind speed and the pressure on the surface [6].

2.3 Case study: Simulation of Tesla Cybertruck

A complete simulation evaluation using a Tesla Cybertruck is provided. The simulation involved increasing the wind speed to 108 km/h and examining the pressure and velocity contours around the car. This case study explores the practical use of simulation results.

2.3.1 Simulation setup

- Vehicle model: Tesla Cybertruck.

- Wind speed: 108 km/h.

- Simulation tools: Ansys.

2.3.2 Results and analysis

- Pressure analysis: the pressure contours showed that the front of the vehicle experienced high-pressure zones, while the rear had lower pressure. This information was used to optimize the placement of the turbine to minimize drag.

- Velocity analysis: the velocity contours indicated areas of high and low wind speed around the vehicle. To maximize energy capture without significantly affecting the vehicle's aerodynamics, the turbine was placed in a stable airflow region.

- Structural analysis: the structural analysis confirmed that the turbine and its mounting system could withstand the aerodynamic forces encountered during operation [6].

3. Experimental validation with Toyota Auris hybrid

3.1 Experimental setup

An experimental validation was performed on the Toyota Auris hybrid, due to its efficient hybrid technology and also aerodynamics. The vehicle is equipped with a roof-mounted wind turbine and other sensors to track its performance.

3.1.1 Test scenarios

The vehicle was tested under various conditions, including speeds, wind directions, and road type. The route of the experiment is illustrated in **Figure 4**.

- Car speeds: tests were performed at 50, 70, 90, 110, and 120 km/h.
- Types of road: the vehicle was tested on both urban and extra-urban roads to simulate different driving conditions.

3.2 Evaluation metrics

The performance of the wind turbines was tested to understand power output, aerodynamic drag, and structural integrity. These measures were also compared to the predictions obtained from the computer simulations to validate the design.

3.2.1 Power output

The electricity produced by the turbine was recorded. The results were then compared to the predictions from the computer simulations in order to validate the design.

- Maximum power output: the turbine generated a maximum power of 110 W at a vehicle speed of 120 km/h.

Figure 4.
The route for the experiment.

- Power output at various speeds: the power output was measured at different vehicle speeds to understand the turbine's performance under various conditions.

3.2.2 Aerodynamic drag

Aerodynamic mapping of the vehicle with and without the turbine. This helped provide insight into how the turbine affects the overall efficiency of the vehicle.

- Effects on fuel consumption: the impact of extra drag on the event's fuel consumption was investigated.

3.2.3 Structural integrity

To ensure that the turbine and its mounting system could withstand forces experienced during operation, the structural integrity of both was assessed.

- Durability testing: the turbine and its mounting system were tested for durability under various driving conditions.

3.3 Case study: Toyota Auris hybrid

In one of the experiments, a Toyota Auris hybrid was utilized to evaluate the performance of a car-mounted wind turbine. The test measured power production and aerodynamic drag at various vehicle speeds. **Figure 5** shows the actual installation of the wind turbine on the car.

3.3.1 Power output

The digital simulations also indicated that at approximately 120 km/h, a peak power of 110 W was produced by the turbine.

3.3.2 Aerodynamic drag

The effect on aerodynamic drag was negligible, with only a minor increase in the drag coefficient. This confirmed the design and placement of the turbine. The tests were performed with and without a wind turbine mounted on the car. **Figure 6** shows that there is a slight difference of approximately 0.1 L/100 km.

Figure 5.
Wind turbine mounted on the car.

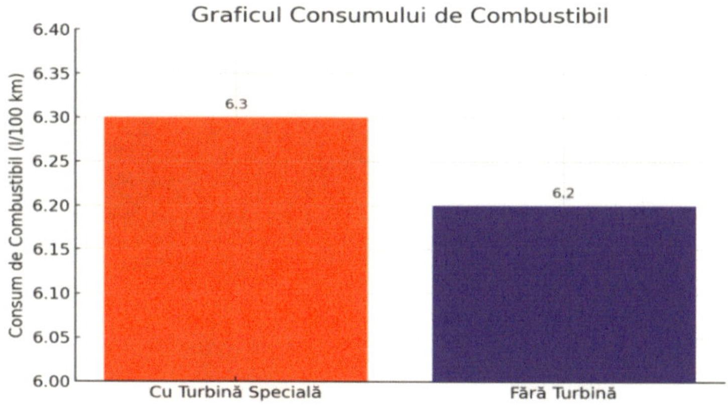

Figure 6.
Fuel consumption.

3.3.3 Fuel consumption

The wind turbine and solar panels reduced fuel consumption by approximately 1 L/100 km. This showcased the potential of the system to enhance the overall vehicle efficiency.

3.4 Integration of solar panels

The solar panels used in the studies were flexible and attached to the vehicle's roof. The integrated impact of wind turbines and solar panels on vehicle performance was investigated, indicating significant improvements in terms of fuel efficiency and range [7]. **Figure 7** illustrates the integration of the hybrid system, including both wind turbines and solar panels, on the vehicle.

3.4.1 Overview of solar panels

- Type: Flexible solar panels.
- Power output: 100 W.
- Mounting method: Magnetic mounting for easy installation and removal.

3.4.2 Performance indicators

- Energy generation: the solar panels generated an additional 1.2 kWh of energy per day under optimal conditions.
- Impact on fuel consumption: the combined system of wind turbines and solar panels reduced the car's fuel consumption by approximately 1 L/100 km.
- Extended range: the additional energy generated by the solar panels extended the vehicle's range by approximately 14 km per day.

Figure 8 compares the fuel consumption of the vehicle with and without the solar panel, highlighting the additional benefits of integrating solar energy.

Figure 7.
Hybrid system integrating on the car [7].

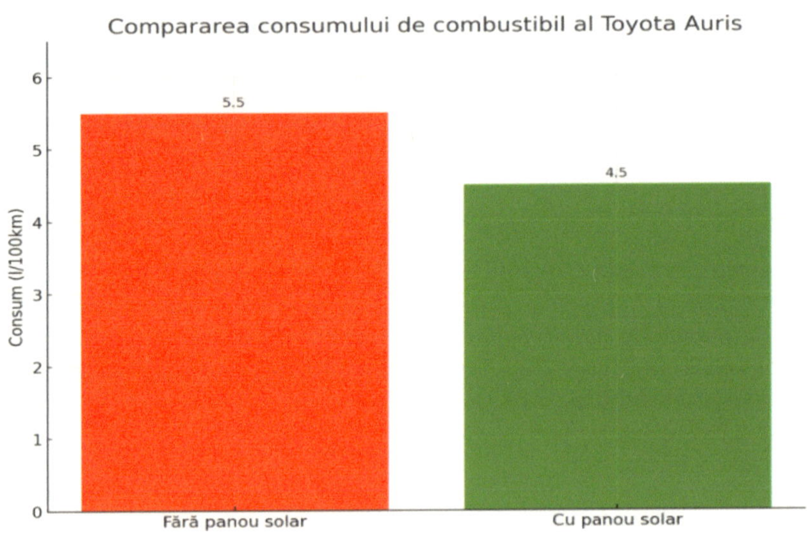

Figure 8.
Fuel consumption with and without solar panel.

3.5 Experimental difficulties and solutions

Many problems were faced during the experiments and solutions were devised to overcome them.

3.5.1 Aerodynamic impact

This may affect fuel consumption and performance as the aerodynamics of the vehicle may be impacted with wind turbines installed. In order to reduce these impacts, the turbines' design and location were extensively studied. Computer simulations guided these decisions, maximizing energy generation and reducing aerodynamic impact.

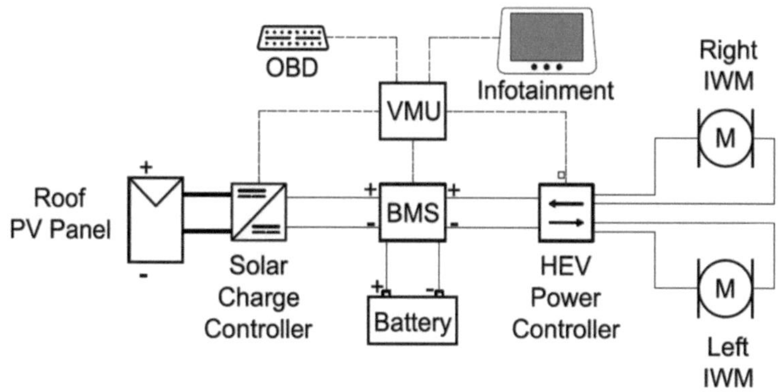

Figure 9.
System integration.

3.5.2 Harnessing wind energy effectively

Capturing wind energy efficiently in a moving car is challenging. This issue was addressed by designing turbines specifically adapted to the conditions experienced by a moving vehicle. The turbines were tested and modified to achieve maximum efficiency. Essentially, the positions of the rotor and stator were swapped.

3.5.3 Safety considerations

Safety is critical. The turbines were built with passengers' safety in mind and set securely in place. They were securely mounted with suction cups, ensuring a firm attachment while allowing for quick removal if needed.

3.5.4 Integration with existing car systems

The integration of the wind energy system in conjunction with the existing systems of the car presented technical challenges. To solve these, the wind energy system was designed to be compatible with normal car propulsion and battery charging technologies, ensuring smooth integration. **Figure 9** presents a block diagram of the connection of the wind energy and solar energy system with the existing car systems, showing how the components are integrated to ensure smooth functionality.

3.5.5 Cost and scalability

The wind energy system has to be scalable to be economically viable. This was achieved by using low-cost components and materials whenever possible and developing the system to be adaptable to different vehicle types.

4. Conclusions

This research thoroughly examines the potential and feasibility of integrating wind and solar power into electric and hybrid vehicles. The primary goal was to determine how these renewable power sources can contribute to increased energy efficiency and sustainability in road transport.

The most significant finding is that using solar and wind power as electric vehicle power sources can substantially cut both greenhouse gas emissions and fossil fuel dependence. This not only helps protect the environment but also increases the range of electric vehicles. The study highlighted that emerging technologies in renewable energy, such as flexible solar panels and miniature wind turbines, can be successfully integrated into vehicle design, allowing for better energy efficiency and extended range.

Nevertheless, significant challenges still remain. High initial prices, the necessity for suitable charging and energy storage infrastructure, and the variability of energy sources, such as a lack of sunlight or wind, are all issues that must be addressed. The study of real-world examples of automobiles powered by solar and wind energy demonstrated that, under the right conditions, these vehicles can be not only practical but also incredibly energy-efficient.

Looking into the future, with technical improvements and growing awareness of environmental issues, it is anticipated that vehicles driven by renewable energies will become the standard in the automobile industry. The initial introduction of these vehicles has the potential to impact the financial situation of the automobile industry. As demand for fossil fuels declines, economies depending on oil and gas may suffer. However, this change provides new prospects for renewable technology research, development, and manufacturing.

Another big advantage of moving to renewable-powered cars is increased energy security. Dependence on oil and gas imports can be decreased, increasing energy security and lowering vulnerability to price changes or supply disruptions. To realize the potential of vehicles driven by renewable energies, collaboration among multiple sectors is required, from vehicle manufacturers and energy suppliers to researchers and legislators.

Even while there are immediate obstacles to adopting cars driven by renewable energy, the long-term prospects are positive. With enough investments in infrastructure, research, and development, electric vehicles driven by renewable energy can become the main form of sustainable transportation in the future. As they become more popular, the incentive for innovations in this space will grow. Electric vehicles driven by renewable energy can be connected with other systems, such as smart energy grids, allowing them to operate as mobile batteries.

The transition to renewable-powered vehicles necessitates not only technological changes but also shifts in public behavior and views. In addition to lowering greenhouse gas emissions, the use of renewable energies in road transportation offers the potential to minimize air pollution, safeguard habitats, and contribute to the preservation of natural resources. A strong regulatory framework and supportive public policies are critical for accelerating the transition to renewable-powered cars.

Finally, the combined efforts of research, development, innovation, and global collaboration can make renewable-powered vehicles the primary mode of transportation in the future. This transformation will not only help to protect the environment and public health, but it will additionally open the path for new views in technical innovation, economic growth, and sustainability.

DOI: http://dx.doi.org/10.5772/intechopen.1009154

Author details

Cristian Helera* and Dan Alexandru Stoichescu
National University of Science and Technology Politehnica Bucharest, Bucharest, Romania

*Address all correspondence to: cristian.helera@stud.etti.upb.ro

References

[1] Bergek A, Berggren C. The impact of environmental policy instruments on innovation: A review of energy and automotive industry studies. Ecological Economics. 2014;**106**:112-123, SSN0921-8009. DOI: 10.1016/j.ecolecon.2014.07.016. Available from: https://www.sciencedirect.com/science/article/pii/S0921800914002183

[2] Habib HUR et al. Analysis of microgrid's operation integrated to renewable energy and electric vehicles in view of multiple demand response programs. IEEE Access. 2022;**10**:7598-7638. DOI: 10.1109/ACCESS.2022.3140587

[3] Ferguson AR. Wind power: Benefits and limitations. In: Pimentel D, editor. Biofuels, Solar and Wind as Renewable Energy Systems. Dordrecht: Springer; 2008. DOI: 10.1007/978-1-4020-8654-0_6

[4] Schütz T. WLTP—On the increased importance of aerodynamics and impact on development procedures. In: Bargende M, Reuss HC, Wiedemann J, editors. 16. Internationales Stuttgarter Symposium. Proceedings. Wiesbaden: Springer; 2016. DOI: 10.1007/978-3-658-13255-2_1

[5] Zhou L, Elemam MA, Agarwal RK, Shi W. Computational fluid dynamic (CFD). In: Discrete Element Method for Multiphase Flows with Biogenic Particles. Cham: Springer; 2024. DOI: 10.1007/978-3-031-67729-8_4

[6] Cristian H, Dan S. Using the car in motion relative wind power for charging the car batteries. In: 2022 IEEE 9th Electronics System-Integration Technology Conference (ESTC), Sibiu, Romania. 2022. pp. 190-193. DOI: 10.1109/ESTC55720.2022.9939499

[7] Cristian H, Alexandru SD. Experimental analysis of using relative wind power in automotive field. In: 2023 13th International Symposium on Advanced Topics in Electrical Engineering (ATEE), Bucharest, Romania. 2023. pp. 1-5. DOI: 10.1109/ATEE58038.2023.10108207

Chapter 5

On the Control Strategy of Small-Scale Wind Systems for Wide Operating Range

Cedric Caruana and Claire Grima

Abstract

Small-scale wind systems have the potential to contribute to the realisation of ambitious decarbonisation targets by taking advantage of their higher energy density and smaller footprint requirements compared to large-scale wind systems. However, marked cost and performance improvements are required for small-scale wind systems to become viable solutions. This chapter focuses on the development and experimental validation of an overall control strategy for a grid-connected small-scale wind-energy conversion system with a nine-bladed rotor and a permanent magnet generator. The control strategy aims to extract maximum power in the variable-speed region and limit power to the rated value at higher wind speeds using a predetermined control characteristic stored as a lookup table. The control characteristic is derived by superimposing the PMG power characteristic on the wind turbine aerodynamic power curves and translating the operating points to the DC link variables using a simplified software model. Experimental tests were conducted under controlled laboratory conditions and in the field using a prototype multi-bladed wind system. The results demonstrate the complexity of the waveforms due to the nonlinearity introduced by the rectifier stage and the tight control of the wind system by the control strategy. The study also reveals the impact of generator temperature rise on the system characteristics and operating point over extended periods of operation. The findings highlight the importance of considering temperature effects in the design and operation of small-scale wind systems.

Keywords: small-scale wind energy, permanent magnet generator, control strategy, maximum power point tracking, soft-stall, field tests, temperature effect

1. Introduction

Wind energy has evolved into a mature technology over recent years, with continuous upscaling of wind turbine power ratings. A global cumulative capacity of approximately 1.4 TW, spread across all continents, was achieved by 2024 [1]. The installed wind power capacity meets over 10% of the global electricity demand [2], with seven countries generating more than 30% of their electricity mix from wind. The lack of space on land and improved wind resources have extended the relevance of wind turbines to offshore sites. However, these statements refer to large-scale wind systems.

There is at least 1.8 GW of small wind capacity globally from over 1 million turbine installations [3]. Dedicated financial incentive programmes led to increased installations in the first decade of the century. In 2017, an estimated 77.46 MW of small-scale wind projects were installed in Italy. As the incentives were dropped, the installation numbers dropped significantly [4]. Recent installations have been driven by China's rural electrification initiatives and India's renewable energy targets [5]. The global small-scale wind market is projected to reach between USD 2–3 million by the year 2030 [6].

Small-scale wind systems (SSWS) have not experienced similar growth owing to several operational challenges. Units installed in urban or built environments have been reported to underperform, failing to reach their expected energy yield [4]. Operating under high wind speeds is challenging, with possible premature system failure [7]. A higher levelised cost of energy (LCOE) compared to alternative renewable energy sources is also an issue [4]. Average capacity factors are shown to include large variability in [4], reflecting the impact of wind resource quality on turbine performance. The same technology can achieve significantly different overall performances at different locations. Large-scale wind systems are generally deployed in locations with high wind resources. In contrast, the main consideration in the siting of SSWS is the proximity to the load. The wind resources at the location often exhibit low average speeds and high turbulence. Furthermore, these systems are typically installed at lower altitudes [7].

Current typical applications include electric power generation for farms and isolated communities as part of hybrid systems, remote monitoring stations, and telecommunication towers [4]. In countries with lower levels of electrification, SSWS are considered viable solutions for providing electricity in rural areas [4]. The current transition to distributed energy sources, combined with the evolution of smart grids, opens up multiple roles for which SSWS can contribute to the realisation of ambitious decarbonisation targets. This opportunity can foster renewed interest in SSWS, which take advantage of their higher energy density and smaller footprint requirements. However, marked cost and performance improvements are required [4].

Small-scale wind systems are mostly based on direct drive shafts that couple the rotor with a compatible permanent magnet generator (PMG) [8]. PMGs offer a simple structure, high efficiency and energy density and ease of control [9]. There are two main electrical power conversion topologies. The first topology comprises a back-to-back converter configuration. The first converter stage regulates the operation of the generator, whereas the second stage facilitates the interface to the grid through an inductive line filter [10]. The second topology replaces the first converter stage with a passive diode rectifier. In both topologies, fully rated power converters are used. The first topology is more flexible and controllable but is more expensive. The second topology is a simpler and lower-cost alternative [9]. Considering also its higher reliability, the second topology is favoured by the industry [11]. A DC-DC converter stage is commonly added between the rectifier and inverter stages to increase the controllability of power conversion [7]. In particular, the voltage at the output of the rectifier stage can be regulated, whereas the voltage at the input of the inverter stage is kept constant [12]. The use of the rectifier stage restricts the control scheme to regulate the generator rotational speed indirectly [9]. A linear relationship between the DC link voltage and rotational speed is then commonly assumed [13].

Smaller-scale wind turbines generally do not exhibit the same level of aerodynamic refinement as large-scale wind turbines [4]. There is no dominant archetype, and multiple designs and layouts are offered in the market. This compounds efforts for design standardisation [4]. However, cost constraints are common design limitations. Large-scale wind turbines utilise yaw mechanisms to actively align the rotor with the mean wind direction. However, such mechanisms are not financially feasible for smaller-scale systems [4]. Instead, some form of a furling mechanism is typically used. Similarly, blade pitch control, which can regulate power at higher wind speeds, is not commonly available in small-scale systems. Attempts to achieve more effective and robust control include moving away from furling towards more active strategies [4].

Small-scale wind turbine manufacturers are typically forced to select from a limited pool of PMG options, leading to marked differences between the generator and blade characteristics [4]. Consequently, it is crucial to define the operating trajectory of the generator to align the operation with the blade's aerodynamic characteristics. A process for the selection of single- and multiple-constant-voltage modes to maintain operation in the high-efficiency region was proposed in [14]. PMG is a flexible electrical machine that allows multiple operating points; however, the operating efficiency varies. Therefore, it is important to maintain the operating trajectory in the high-efficiency region of the generator. Additionally, PMGs provide a significant overload capability, which facilitates electrical braking. A key indicator of the PMG overload capability is the low-load angle under rated conditions [15].

This chapter develops an overall control strategy for a multi-bladed SSWS. An experimental testing is carried out under laboratory-controlled conditions and in the field. The nonlinearity introduced by the rectifier stage is characterised. Selected characteristics against the DC link voltage and the rotational speed are extracted to examine the behaviour of the electrical system. The effect of generator temperature is revealed, and its influence on the system behaviour is analysed. The rest of the chapter is organised as follows. Section 2 reviews published control strategies extending up to high-wind-speed conditions. Section 3 describes the proposed process for the development of the control characteristic. Section 4 presents the extracted characteristics from experimental tests. Section 5 concludes the chapter.

2. Review of control strategies

Wind turbines transform the kinetic energy of moving air into mechanical energy in the rotor [11]. The amount of extracted energy depends not only on the strength of the wind resource but also on the control strategy in use [16, 17]. For operation below the rated wind speed, the rotor speed is varied to track the wind speed, aiming to maintain the operation within a small interval around the peak point of the power coefficient characteristic [16, 18]. This objective translates to a distinct operating point for each wind speed up to the rated value [12]. At higher wind speeds, the control should confine the generator power and rotational speed within safe levels to protect the system from damage. Given the lack of aerodynamic control, regulation must be performed through electrical control [19]. The imposition complicates the overall power control strategy, particularly at higher wind speeds [8]. Passive stall mechanisms may be embedded in the blade design; however, they can increase the turbine fatigue load and add complexity to the design [20]. Soft-stall strategies are promising alternatives.

2.1 Maximum power point tracking techniques

Multiple maximum power point tracking (MPPT) techniques have been proposed in the literature for operation up to the rated wind speed, each offering different strengths and drawbacks. In this chapter, the techniques are broadly classified into seven categories: Tip Speed Ratio Control; Perturb and Observe; Gradient Approximation; Control Law, Pre-Determined Characteristic, Constant Voltage, Metaheuristic Techniques and Combination of Techniques. Each technique exhibits its own characteristics with respect to implementation complexity, tracking speed, accuracy and sensors required [7]. The different categories are described below.

The Tip Speed Ratio control approach can be described as the most direct approach for operating a wind turbine at the maximum power point. It consists of measuring the wind speed and regulating the rotational speed to maintain the optimal tip speed ratio. This approach is easy to understand but difficult to achieve in practice [21]. Accurate wind speed measurements are difficult and require the use of relatively expensive anemometers and mounting arrangements [12]. Owing to the interaction between the rotor and wind, placing the anemometer on the nacelle leads to a discrepancy between the measurement and wind speed intercepted by the wind turbine [17]. An alternative approach is to predict wind speed using historical data. Autoregressive statistical models driven by the measured rotational frequency were proposed in [16]. Offline-trained neural networks, with rotational speed and power as inputs, were considered in [20]. Multiple approaches have been presented for processing the obtained wind speed information. In [16], the measurement was converted to the rotational speed and used to identify the maximum power reference. The difference between the reference and measured powers is then used to determine the operating DC voltage reference.

The Perturb and Observe technique is alternatively referred to as the Hill-Climb Technique, reflecting hill-shaped power curves. An optimum relationship between the electrical power and another system variable is used to search for and track the maximum power point [9]. The maximisation variable is perturbed, and the resulting power is observed. Based on the power variation with the introduced perturbation, the next step direction and possibly the size are determined [9, 22]. The approach has the advantages of simple implementation, does not require wind speed measurements, and does not require detailed knowledge of system parameters. However, the response is slow towards rapid wind speed fluctuations [11, 23], and the tracking performance depends on the used step size [9, 17]. Improvements to the approach consider the use of variable step sizes to ease the trade-off between the tracking speed and control efficiency [17]. The use of sinusoidal perturbation rather than step perturbation has been proposed in the literature, but it was reported to not provide a distinct advantage over the traditional approach [17]. The DC link voltage slope information was used to enhance the tracking speed and the stability of the algorithm in [9]. Fuzzy-logic-based techniques can be seen to extend the Perturb and Observe approach, where the control action is governed by a set of rules according to the state of the system at that instant. Fuzzy logic control may require the use of speed sensors.

The Gradient Approximation approach works by continuously examining the rate of change of power, with respect to the change in the rotor speed. A zero gradient identifies the maximum power point [23]. This approach has advantages and drawbacks similar to those of the Perturb and Observe technique.

The Control Law approach is based on the identification of a law relating the maximum power condition to a control variable to define an optimal power curve. The control variable can be the rotational speed or DC link voltage. The reference value of the control variable is identified by measuring the extracted power. Most of the reported Control Law strategies rely on the measurement of electrical power, which is equal to the aerodynamic power in the steady state. Because the electrical power cannot instantaneously change with the wind speed owing to the significant inertia of the turbine rotor, the use of aerodynamic power was proposed in [8] to improve the dynamics. As the latter cannot be measured directly, an aerodynamic power observer was proposed. The main disadvantage of the Control Law approach is the required knowledge of the system parameters. In addition to not being accurately known, the parameters can change with the operating conditions and over the lifetime of the wind turbine. Improvements to the technique include the introduction of a robust variable-structure observer to address the difference between the mechanical and electrical torques of the generator [19]. Furthermore, linear and nonlinear state-space techniques have been applied to provide robustness against disturbances [17].

Rather than a control law, a predetermined relationship or characteristic can be used to direct the operation of the MPPT technique. The characteristic is generally stored as a one-dimensional lookup table with one input for the measured variable and one output for the reference variable [16]. In this open-loop control scheme, the controller does not require a mathematical model of the system. Multiple predetermined relationships have been proposed in the literature. The proposed relationships include electrical power versus DC link voltage [16], electrical frequency versus DC link voltage [12], rotational speed versus DC link voltage [11], grid-side power versus frequency [24], electrical power versus rotational speed [17] and DC current versus DC link voltage [10]. The information stored in the lookup tables is based on the averaged values of the two variables [10]. The characteristic is sometimes simplified using approximated straight-line segments [10, 12, 25]. A frequency derivative and a constant offset were added to the measured dc power in [16] to improve the sensitivity and prevent the wind turbine from operating in the stall region. Generator losses were considered in [10]. Similar to the Control Law approach, the predetermined characteristics are assumed to remain constant despite possible variations [17].

A simpler technique that aligns well with off-grid applications with batteries is the Constant Voltage technique [14]. As the name implies, the DC link voltage after the rectifier stage is maintained constant. The effort here is to identify the optimal voltage value. The optimal voltage can be selected based on the wind speed resource variation at the site, for example, through knowledge of the Weibull distribution [16]. The advantage of this approach is its simplicity and lower hardware requirements. However, the power conversion efficiency is reduced at wind speeds below and above the wind speed corresponding to the set voltage [16].

More recent approaches focus on the use of metaheuristic techniques, such as particle swarm optimisation and ant colony optimisation [7, 26]. Both are inspired by the collective behaviour of social organisms to find optimal paths. These algorithms operate using a population of simple agents that interact and share information to explore the search space and converge towards optimal or near-optimal solutions. A rotor speed controller based on fuzzy logic concepts for a small-scale wind turbine with pitch control was proposed in [26, 27].

The last category considered in this chapter is the Combination of Techniques, where two or more of the former techniques are combined to offset the identified drawbacks. In Refs. [17, 21], the Control Law approach is combined with the Perturb and Observe technique. The latter is used to determine/tune the control law coefficient, thus accounting for the influence of external parameters. Thus, the accuracy of the control law is increased.

2.2 Operation above rated wind speed

Several alternative approaches are available for controlling small-scale wind turbine operations under high-wind-speed conditions. These approaches can generally be divided into mechanical and electrical mechanisms [7]. Mechanical braking devices can be used to stop the wind turbine operation when the wind speed exceeds the rated value. Apart from the additional expense of the brakes, the system efficiency suffers because no power is extracted at higher wind speeds [7]. Passive furling mechanisms are alternative mechanical approaches which drive the wind turbine rotor out of the wind when the wind speed exceeds the rated speed. Similar to the mechanical brake solution, the system efficiency decreases. The traditional electrical approach involves dissipative methods. Braking resistors are integrated into the system to limit the rotational speed. The DC link voltage is typically used to detect shaft overspeed conditions and trigger electrical brake operations. The excess power is dissipated in the dump load, whereas the rated power is delivered to the grid. A major disadvantage of electrical braking is that a high power is required from the generator during braking, which may affect its lifetime [7].

Recently, electrical methods in the form of soft-stall control techniques have become popular additions to MPPT algorithms for formulating the overall control strategies [9]. Continued wind-energy capture at higher wind speeds improves system efficiency [7]. Soft-stall techniques are distinguished from the passive stall, where the blade design induces turbulence at higher wind speeds to reduce the lift force, thus limiting power extraction [26]. In the soft-stall approach, the DC link voltage can be fixed to an upper limit to achieve a constant-speed stall control [8]. The voltage and rotational speed will remain constant; however, the output power would still increase with increasing wind speed, although at a lower rate than if MPPT were maintained. The generator and power conversion stages must be rated accordingly [9, 20]. To safeguard the wind system components, the aerodynamic power of the turbine must be constrained to the rated power.

There are two possible ways to reduce the power coefficient when operating at higher wind speeds. The first is to increase the rotational speed above the nominal value to convert excessive wind energy into kinetic energy [9, 19]. The second approach is to reduce the rotational speed below the nominal value. In the first approach, the induced voltage in the generator, and hence the DC link voltage, continues to increase with the rotational speed [28] and needs to be limited. For topologies with a diode rectifier, an external means of limiting the voltage is required. Additionally, the generator requires a sufficiently high maximum speed rating to extend the constant power region. The second approach alleviates the concerns regarding high voltage, allowing higher wind speeds to be reached. However, the generator must be capable of delivering higher torques than those required at the nominal wind speed. This requires a generator with a higher-rated power than the wind turbine rotor [19].

In the second approach, the turbine is forced into the constant power stall region by decreasing the DC link voltage from the nominal value at higher wind speeds. The shift in the operating point increases the generator torque, forcing the turbine to stall [8]. The SSWS will then be able to extend its high-wind-speed withstand capability and cut-off speed [10, 18]. The energy production is increased by the extending the operating range at the rated power output. Different techniques can be applied to extend the constant-power speed range. Reference [28] proposed increasing the generator stator inductance by introducing additional cores enclosing the end windings. In contrast to operation in the MPPT region, steady-state operation in the stall region is unstable. Hence, additional control terms are required to stabilise system operation [9, 18]. A customised controller design is necessary to comply with the characteristics of individual wind turbines [20].

3. Development of control characteristic

This section starts by presenting a background theory related to SSWS operation. The considered multi-bladed system is described. A process for the derivation of an overall control characteristic, based on the described system, is then presented.

3.1 Background theory

The fundamental relationship between the captured power and the wind speed in a variable speed wind turbine is given by the nonlinear equation expressed in Eq. (1). The power coefficient C_p reflects the rotor's aerodynamic properties. Its value varies as a function of the tip speed ratio λ and the pitch angle β.

$$P_{wt} = \frac{1}{2}\rho A C_p(\lambda, \beta) V_w^3 \tag{1}$$

Small-scale wind turbines are generally of fixed pitch; therefore, the relationship in Eq. (1) can be simplified to Eq. (2). The tip speed ratio λ expresses the variation of the rotor tip speed with respect to the wind speed, as given by Eq. (3).

$$P_{wt} = \frac{1}{2}\rho A C_p(\lambda) V_w^3 \tag{2}$$

$$\lambda = \frac{\omega_{wt} R}{V_w} \tag{3}$$

PMGs are widely used in SSWS because of their flexibility and overload capability. Considering a dq frame orientated to the permanent magnet flux, the stator voltage components of a surface-mount PMG operating under steady-state conditions are expressed by Eqs. (4) and (5). The developed generator torque τ_g is expressed by Eq. (6) in terms of the i_{sq} current component.

$$v_{sd} = R_s i_{sd} - L_s \omega_e i_{sq} \tag{4}$$

$$v_{sq} = R_s i_{sq} + \omega_e \left(L_s i_{sd} + \psi_{pm} \right) \tag{5}$$

$$\tau_g = \frac{3}{2} p \psi_{pm} i_{sq} \tag{6}$$

The generator input power P_{gi} can be determined using the electromagnetic torque and the rotational speed n_g, as expressed in Eq. (7). Neglecting the gearbox losses, the difference between P_{wt} and P_{gi} determines the rotational speed, as indicated in Eq. (8). Under steady-state conditions, the two powers are equal.

$$P_{gi} = \tau_g \omega_m \tag{7}$$

$$P_{wt} - P_{gi} = \omega_m J \frac{d\omega_m}{dt} \tag{8}$$

The active and reactive powers at the output of the generator can be expressed in terms of the dq frame voltage and current components, as expressed in Eqs. (9) and (10). The rectifier stage after the PMG imposes a unity power factor. This condition forces a relationship between the current components [29], as indicated by Eq. (11). The implication of Eq. (11) is that the PMG i_{sd} current component follows the i_{sq} component as it varies with the wind. The i_{sd} variation leads to a field-weakening effect, which is however not controllable.

$$P_{go} = \frac{3}{2}\left(v_{sd} i_{sd} + v_{sq} i_{sq}\right) \tag{9}$$

$$Q_{go} = \frac{3}{2}\left(v_{sd} i_{sq} - v_{sq} i_{sd}\right) \tag{10}$$

$$L_s\left(i_{sd}^2 + i_{sq}^2\right) + \psi_{pm} i_{sd} = 0 \tag{11}$$

3.2 Power curve operating regions

A wind turbine power curve generally exhibits four operating regions, as illustrated in **Figure 1**. In region I, the wind turbine does not operate because of insufficient wind speed. In region II, which spans from the cut-in wind speed to the rated wind speed, the wind turbine is operated at a variable speed to track the maximum power from the incident wind. Numerous control strategies for the operation in this region were reviewed in Section 2. Their applicability depends on the turbine design, matching between the wind turbine rotor and the generator, and power conversion topology. As the wind speed increases, the operation enters region III, where the extracted power is limited to protect the generator and the power converters. Different soft-stall

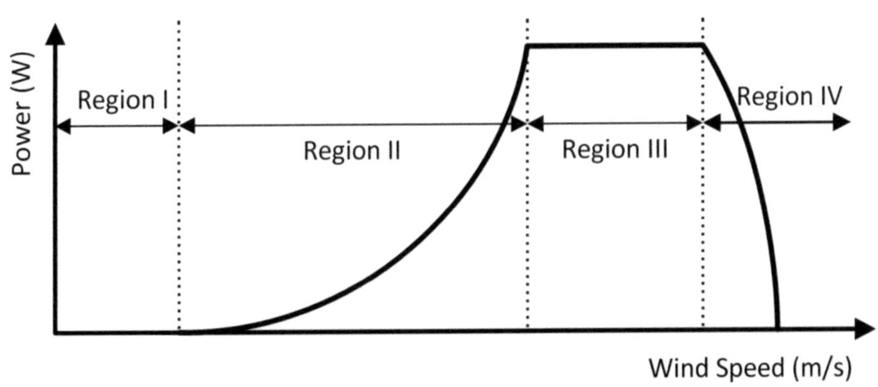

Figure 1.
Typical small-scale wind turbine power curve with four operating regions.

mechanisms have been employed for this purpose. As the wind speed increases further, electrical braking is typically used to limit the rotational speeds. At even higher wind speeds, a form of furling mechanism kicks in to move the turbine rotor out of the wind.

3.3 System description

This chapter considers a grid-connected SSWS with a nine-bladed rotor. The choice of rotor addresses the low-capacity factor concerns by allowing for improved energy extraction at locations with lower average wind speeds. A power conversion topology with a passive rectifier is considered, where a commercial wind inverter is used to implement the various stages. A schematic representation of the system setup, which identifies the variables at different power conversion stages, is shown in **Figure 2**. Because of the use of a commercial PMG with a multi-bladed rotor, a small gearbox was installed between the rotor and the PMG to improve the matching between the aerodynamic and electrical parts.

Mathematical modelling of the rotor performance was conducted in [30] using blade element momentum theory. The resulting variation of the power coefficient C_p with the tip speed ratio under the steady-state conditions for a fixed pitch angle is shown in **Figure 3**. The maximum predicted power coefficient was 0.426 at a tip

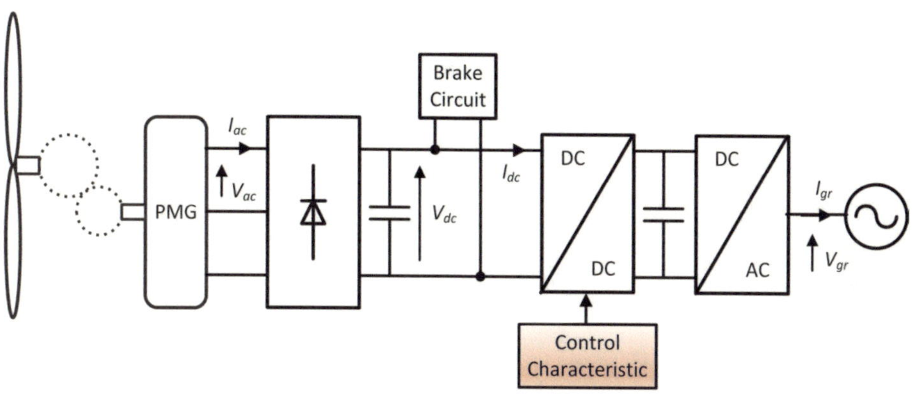

Figure 2.
Schematic of the multi-blade SSWS setup. The relevant variables at different power conversion stages are identified. The variables are defined in the Nomenclature section.

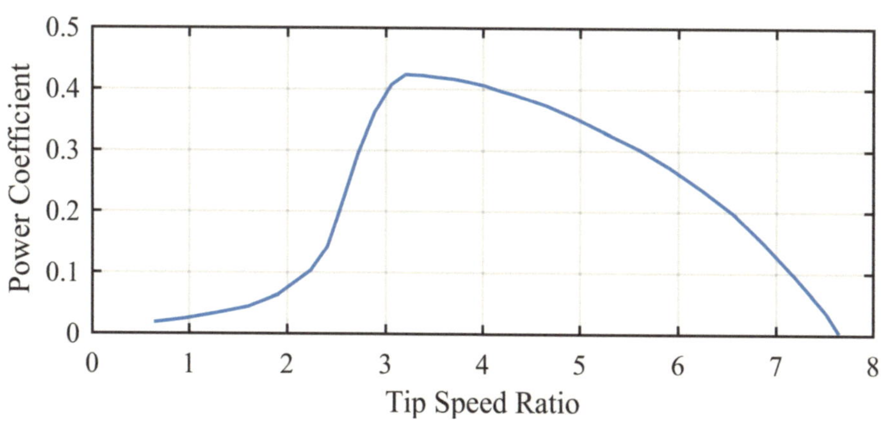

Figure 3.
Power coefficient C_p versus tip speed ratio λ characteristic of the multi-bladed rotor.

speed ratio of 3.2. A lower value of the optimal tip speed ratio reflects the solidity of the rotor. The salient parameters of the wind turbine are presented in **Table 1**.

A predetermined characteristic is used to implement the control methodology. The characteristic is stored as a one-dimensional lookup table, relating the grid-side power P_{gr} to the DC link voltage V_{dc}. Based on the measured power, it provides a voltage reference for the DC-DC converter stage, as illustrated in **Figure 2**. The relevant generator and converter parameters are listed in **Table 2**. The GCI-2K inverter accommodates a 40-point control characteristic, which allows for a more accurate representation of the control trajectory than the use of straight-line approximations. However, the wide DC input voltage range limits the number of points which can be used. A lookup table assumes a deterministic relationship between the input and output variables. Therefore, only one corresponding output value can be stored for any given input value. This requirement limits the choice of operation at wind speeds higher than the rated speed. The available option is to increase the rotational speed of the turbine above the nominal value to reduce the power coefficient and, hence, the extracted power. To safeguard the system against excessive rotational speed, the brake circuit is triggered once the DC link voltage V_{dc} reaches the maximum value.

3.4 Process for development of control characteristic

The control characteristic covers operation in Regions II and III of the power curve shown in **Figure 1**. As identified in Section 2, the objective of the control characteristic is to extract the maximum possible power in region II and limit the power to the

Parameter	Value
Number of blades	9
Rotor radius (m)	1.7
Moment of inertia (kgm^2)	22
Rotor solidity (%)	21
Maximum power coefficient	0.43
Optimal tip speed ratio	3.2
Gearbox ratio	1.46

Table 1.
Wind turbine parameters.

Permanent magnet generator		Wind inverter	
PMG model	**GL-PMG-1800**	**Inverter model**	**GCI-2K**
Rated power (kW)	1.8	Rated power (kW)	2
Rated speed (rpm)	480	DC voltage input range (V)	30...750
Maximum speed (rpm)	600	Characteristic points	40
Pole pairs	8	Controller model	GCB-5K
Phase resistance (Ω)	5.0	Dump load (kW)	4

Table 2.
PMG and wind inverter parameters.

rated value in region III. In this chapter, a multi-bladed rotor is considered to improve the energy extraction at lower wind speeds. The rotor was oversized with respect to the generator to further reduce the cut-in wind speed. This design choice necessitates a change in the control philosophy of region II.

3.4.1 Region II

The C_p-λ curve shown in **Figure 3** can be expanded into a family of curves showing the wind turbine power P_{wt} conversion capability with respect to the rotor rotational speed at different wind speeds. Given the integration of the gearbox in the multi-bladed system, as illustrated in **Figure 2**, the rotational speeds are referred to the generator side. The rotational speed is given by Eq. (12). For each wind speed, the power and rotational speed coordinates are obtained from Eqs. (2) and (12) by varying the λ value in the applicable range. The resulting family of curves for wind speeds up to 10 m/s is shown by the blue traces in **Figure 4**.

$$n_g = \frac{60}{2\pi}\frac{\lambda V_w}{R}G \tag{12}$$

To maximise the performance, MPPT is generally used, where the maximum power is determined by the rotor's aerodynamic characteristics. This criterion cannot be followed in this case because of the use of a lower-rated generator. The objective of maintaining a high C_p value remains unchanged. Because the operating criterion is detached from the wind turbine characteristics, the onus falls on the generator. PMGs are flexible machines capable of operating at multiple points. A technique similar to that proposed in [14] can be used, where multiple constant voltage modes are examined to identify the optimal operating voltage for each wind speed. As identified in [14], it is important to maintain the selected operating points in the high-efficiency region of the generator. The generator's copper losses are mainly considered because the iron losses are not straightforward to determine. However, following this process still leaves some operating point choices to be considered. Thus, an alternative approach is proposed.

Considering the lower-rated PMG, a suitable starting point for the derivation of the control trajectory is the generator's torque characteristic provided by the manufacturer. It is typically expressed against the generator rotational speed. It can be considered to set an upper boundary because operation at lower torques leads to increased efficiency. The torque characteristic can be converted into a generator input power P_{gi} characteristic by multiplying the operating points by the respective rotational speed. The resulting P_{gi} versus the rotational speed characteristic can then be superimposed on the family of wind turbine aerodynamic power curves. A 10% torque reduction was introduced in the process to allow for any overshoot. This derating factor is arbitrary and can be modified.

The superimposed P_{gi} characteristic on the wind turbine P_{wt} curves is shown in **Figure 4**. The intersection between the P_{gi} and P_{wt} curves at each wind speed identifies suitable steady-state operating points for variable-speed operation in region II. Depending on the particular shape of the C_p-λ curve, at lower wind speeds, slight adjustment may be required to prevent the intersection of the control trajectory with the stall region sides of the higher wind speed traces. If this occurs, the operation will be stuck at this point. As the wind speed varies, the applicable P_{wt} trace changes immediately, but the rotational speed varies slowly, mainly because of the

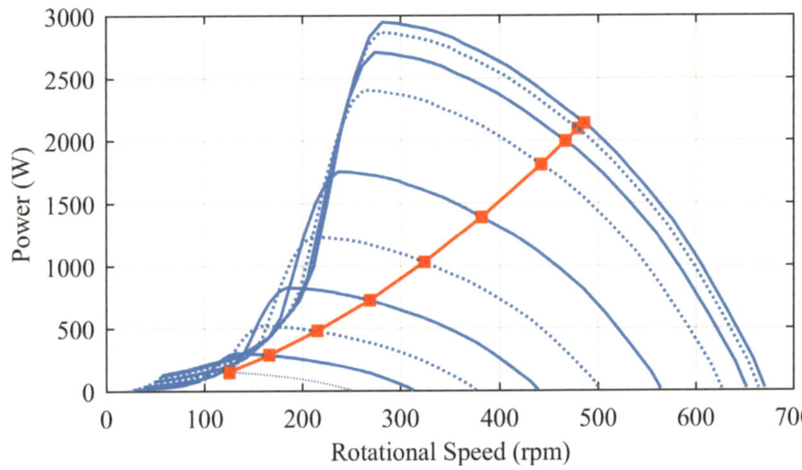

Figure 4.
Identification of the steady-state operating points in region II of the control characteristics. Considered wind speeds: 4–10, 10.4, 10.6 and 10.7 m/s.

rotor inertia. In reality, the operation does not settle at a particular point because the wind speed continuously changes. Thus, a set of quasi-steady-state operating points can be identified, where each *j*th point is characterised by a (V_w^j, n_g^j, and P_{gi}^j) set. The obtained values cannot be directly programmed into the predetermined control characteristic.

Because the inverter characteristics typically use the DC link voltage V_{dc} as a reference, the next step is to translate the operating point P_{gi}^j and n_g^j variables to the DC link variables P_{dc}^j and V_{dc}^j. However, as identified in the previous section, the rectifier stage introduces nonlinearity into the system. To avoid deriving complex mathematical equations, a simplified software model was developed to perform the translation. The simplified model considers the PMG, rectifier, and variable dc source, as shown in **Figure 5**. The PMG is operated at the extracted rotational speed n_g^j. The DC link voltage V_{dc} is varied until P_{gi} coincides with the extracted value. Once the condition is achieved, the applied V_{dc} and corresponding DC link variables are saved. Additional information, such as the generator torque and current components, can also be extracted. The variable set for each operating point is thus extended to (V_w^j, n_g^j, P_{wt}^j, V_{dc}^j, P_{dc}^j, τ_g^j, i_{sq}^j, and i_{sd}^j). The new variable set allows plotting the P_{dc} vs. V_{dc} characteristic, as shown in region II part of the blue trace in **Figure 6**.

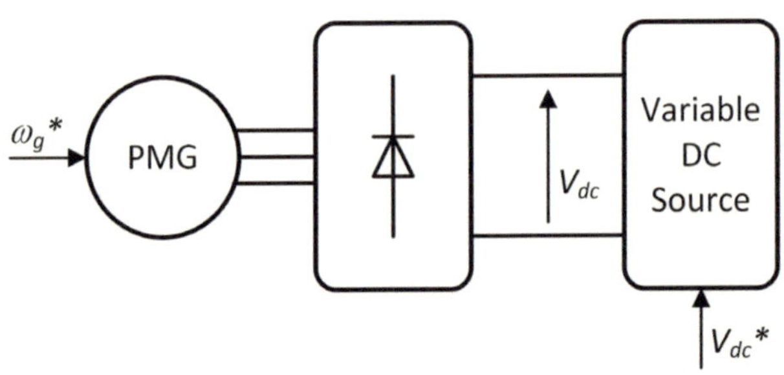

Figure 5.
Simplified software model for the translation of operating variables.

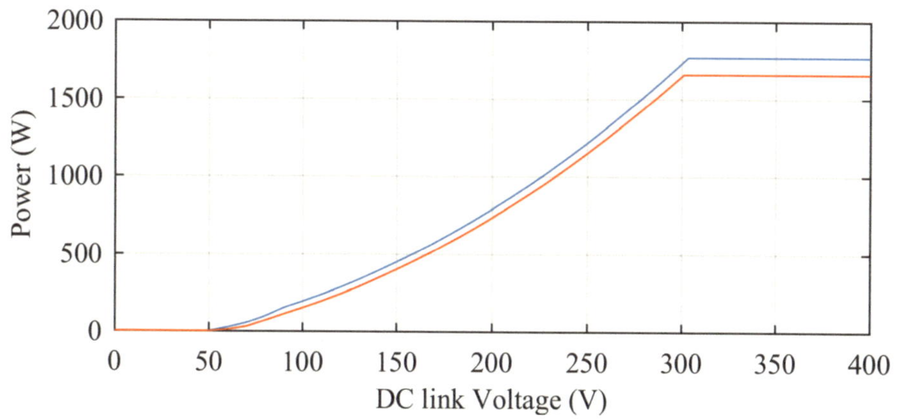

Figure 6.
Derived P_{gr} *and* P_{dc} *vs.* V_{dc} *control characteristics. Legend:* P_{dc}*—blue trace,* P_{gr}*—red trace.*

3.4.2 Region III

The objective of the control characteristic in this region is to limit power extraction to safeguard the system. Two methods for reducing the power coefficient C_p, and hence the extracted power at higher wind speeds, have been previously identified. The first approach is to reduce the rotational speed to operate below the nominal value, thus soft-stalling the turbine rotor. The second approach is to increase the rotational speed to operate above the nominal value. As noted earlier, the use of a lookup table restricts the choice to the second approach only. Moreover, the lower-rated generator does not have the capability to produce the required torque for soft-stall operation. By neglecting the rectifier stage losses, the generator-rated power sets the constant P_{dc} value. The span of region III is determined by the maximum speed rating

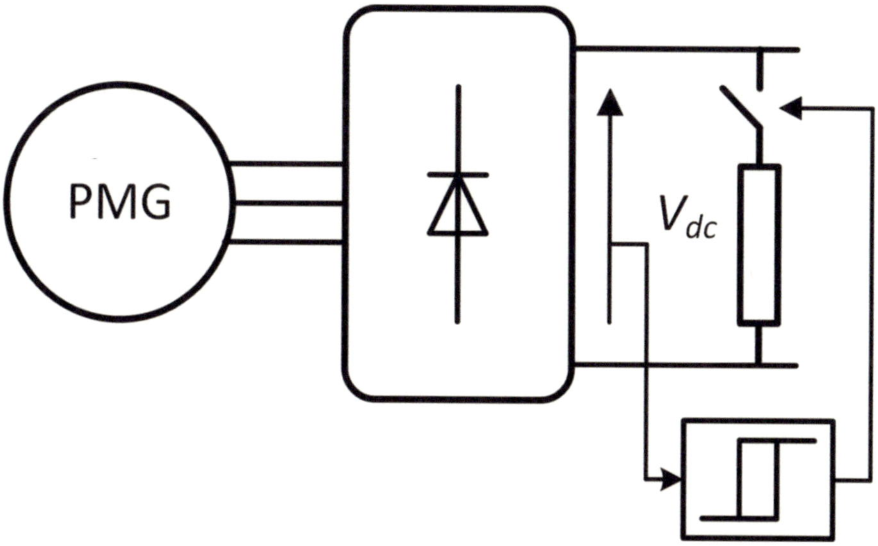

Figure 7.
Hysteresis controller for electrical brake circuit.

of the generator. A sample of the resulting P_{dc} vs. V_{dc} characteristic covering both regions II and III is illustrated by the blue trace in **Figure 6**.

The last step in the characteristic derivation process involves expressing the power in terms of P_{gr} rather than P_{dc}. The inverter efficiency plays a prominent role in this process. The efficiency characteristic is generally published by the manufacturer. If not fully specified, experimental tests can be conducted to identify the missing information. The resulting P_{gr} and P_{dc} characteristics are depicted in **Figure 6**.

3.4.3 Overspeed protection by electrical brake

The electrical brake circuit introduces a dump-resistive load across the DC link. The resistive load is activated by a switch that is typically controlled by a hysteresis controller, as shown in **Figure 7**. The hysteresis controller prevents the chattering of the switch. A DC link voltage related to the generator's maximum rotational speed can be used as the upper threshold to trigger the brake operation. A lower threshold is then used to deactivate the switch as the rotational speed decreases owing to the additional electromagnetic torque.

4. Experimental tests

Experimental tests were first performed under controlled laboratory conditions. The results identify the complexity of the waveforms owing to the nonlinearity introduced by the rectifier stage. The control characteristic was then implemented on a prototype multi-bladed SSWS and tested in the field. The resulting system performance was studied.

4.1 Controlled laboratory tests

A control characteristic, similar to that shown in **Figure 6**, was first programmed on a wind emulator test rig, which was operated under controlled conditions in a laboratory. A schematic representation of the wind-emulator rig is shown in **Figure 8**. It consists of a PMG coupled to a DC motor, with a gearbox in between to match the rotational speeds. The DC motor is driven by a DC drive, which accepts torque and speed references. The PMG and the power conversion topology are similar to the schematic of **Figure 2**.

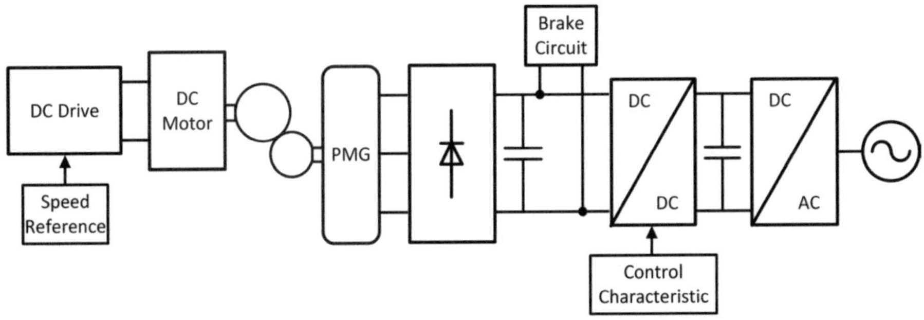

Figure 8.
Schematic of wind-emulator test rig.

The tests were conducted with the PMG driven at constant rotational speeds corresponding to the steady-state operating points spread along the control trajectory. Owing to the rating of the DC drive, the maximum power of the control characteristics had to be limited to 1.45 kW.

The captured waveforms at a rotational speed of 350 rpm, which falls within region II of the control characteristic, are shown in **Figure 9**. The traces were sampled at a very high frequency using an oscilloscope to portray the full nature of the instantaneous waveforms. **Figure 9(a)** shows the phase current (upper plot) and line voltage (lower plot) on the AC side, at the output of the rectifier. **Figure 9(b)** and **(c)** show the corresponding current and voltage traces at the DC and grid sides, respectively.

The non-sinusoidal nature of the generator current and voltage waveforms is evident. The phase current is discontinuous, with the rectifier operating in mode 2/3 [31], where two diodes operate continuously and three devices operate during current commutation. The line voltage waveform exhibits distinct behaviour during the respective parts of the cycle. This ranges from zero voltage to sinusoidal variation and to being clamped at a constant value. Zero voltage occurs at the periods when the current commutates between the respective phases of the line voltage waveform. The voltage follows a sinusoidal variation during the discontinuous part of the current in the two respective phases, that is, when the corresponding diodes are not conducting. The waveform also exhibits a repetitive voltage transient before entering the sinusoidal phase. When both phases conduct, the instantaneous voltage is clamped to the DC link voltage. The latter is shown in **Figure 9(b)** to be well regulated to the reference value by the DC-DC converter. The DC link current corresponds to the peak of the phase currents and clearly shows the 6-pulse ripple characteristic of full-bridge rectifier operation. The grid-side waveforms are decoupled from the AC-side waveforms. The former waveforms are at a 50-Hz grid frequency. The inverter also operates very close to a unity power factor, which is confirmed by the current being in phase with the voltage.

As the rotational speed increases, the operation moves to the constant power region of the control characteristic. The current and voltage waveforms corresponding to a rotational speed of 500 rpm are shown in **Figure 10**. These are presented in a manner similar to that in **Figure 9**. The phase current waveform shown in

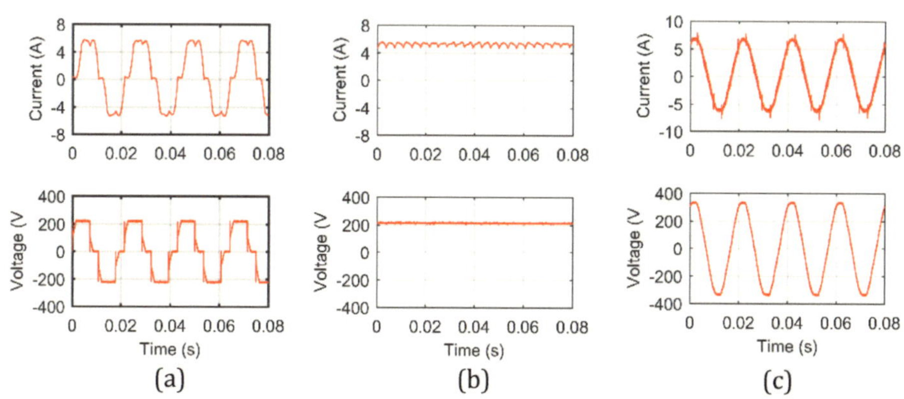

Figure 9.
Emulator-test-rig operation at 350 rpm (region II) showing the current (upper plots) and voltage (lower plots) at: (a) the AC side, (b) the DC side and (c) the grid side.

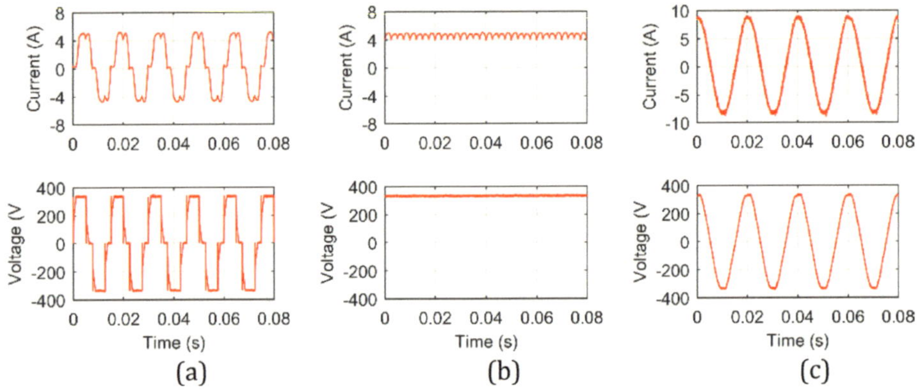

Figure 10.
Emulator-test-rig operation at 500 rpm (region III) showing the current (upper plots) and voltage (lower plots) at: (a) the AC side, (b) the DC side and (c) the grid side.

Figure 10(a) shows a similar waveform to that in **Figure 9**, but with a slightly shorter discontinuous period. The current also reached a lower peak. Similarly, the line voltage V_{ac} exhibits a similar behaviour but reaches a higher voltage level. The DC link waveforms correspond to the peaks of the current and voltage waveforms, as in the previous case. The lower peak current reflects the higher voltage in the constant power region. The grid-side waveforms also show a similar behaviour with increased current amplitude owing to the higher power transfer.

The instantaneous waveforms exhibited significant complexity corresponding to the operation of the devices comprising the power converter topology. However, extracting this information requires processing significant amounts of data. Because the quasi-steady-state operation of the system shows significantly slower dynamics, averaged values are used for subsequent analysis.

4.2 Field tests

The nine-bladed SSWS prototype shown in **Figure 11** was used to perform a series of field tests. The same PMG and power conversion stages as those for the wind-emulator rig were used. A control characteristic similar to the one shown in **Figure 6** was programmed in the wind inverter. Parameters from the AC, DC and grid sides, as defined in **Figure 2**, were sampled at a high frequency using a data acquisition card and averaged over a 1-s interval.

The plotted test data were collected over the three test days with varying wind speeds. The control characteristics used on the first day were slightly modified from those of the subsequent days to obtain temperature matching [32]. For the purposes of this chapter, the modification to the characteristic is not significant and does not affect the subsequent discussions. The parameter variation over time and extracted characteristics with respect to the DC link voltage V_{dc} and generator rotational speed n_g were examined. Multiple runs were performed each day to gauge the operation over extended periods.

4.2.1 Time-domain plots

The wind speed and extracted power variation with time for the first test day are shown in **Figure 12**. **Figure 12(a)** shows that the wind was gusty, reaching peaks of 10–11 m/s. The corresponding P_{dc} and P_{gr} behaviours over time are shown in

Figure 11.
The nine-bladed SSWS prototype used for the field tests.

Figure 12(b). It can be observed that owing to the lower wind speeds, the system operation did not exceed region II of the control characteristic.

Higher wind speeds were present on the second test day, as shown in **Figure 13(a)**. The wind speed was also gusty, with peaks reaching 14–16 m/s. The system operation spreads over the entire control characteristic range. This is evidenced by the flat parts of the power waveforms in **Figure 13(b)**, which corresponds to the operation in region III. Furthermore, the dips during these periods indicate the operation of the brake circuit owing to the higher wind speeds. Higher brake powers are not visible because the DC link current is measured after the brake circuit connection point. The wind conditions for test day 3 were between those of the two previous days.

4.2.2 Extracted characteristics

Following the examination of the time-domain waveforms, several characteristics were plotted to examine the behaviour of the salient system parameters. In this section, data from test day 2 are used because of the higher wind speeds. The characteristics are plotted against the DC link voltage V_{dc} and generator rotational speed n_g.

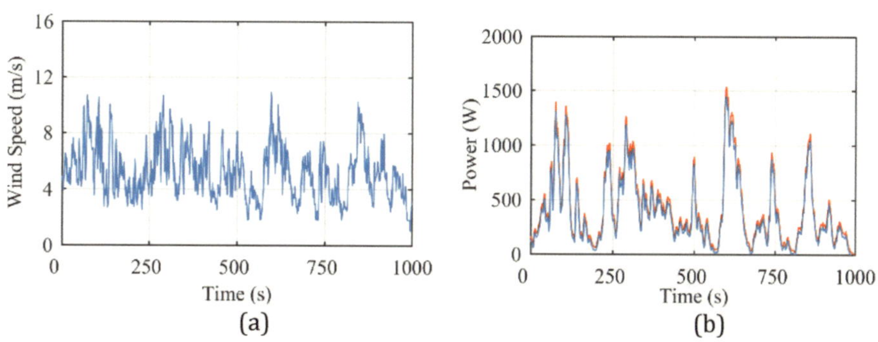

Figure 12.
Conditions for test day 1: (a) wind speed and (b) DC link power P_{dc} *and grid-side power* P_{gr} *variation with time. Legend:* P_{gr}*—blue trace,* P_{dc}*—red trace.*

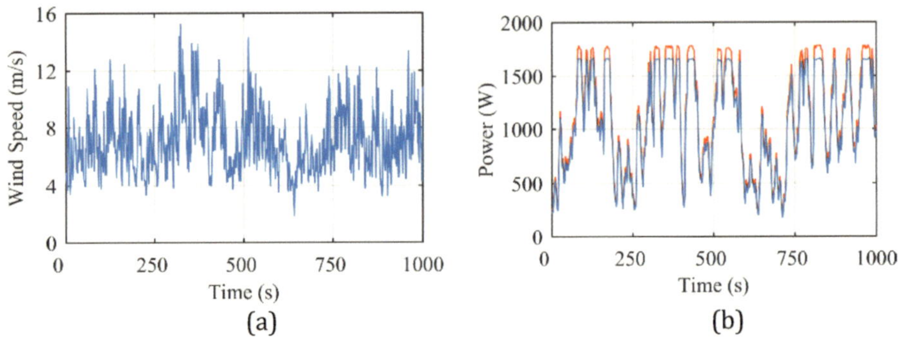

Figure 13.
Conditions for test day 2: (a) wind speed and (b) DC link power P_{dc} *and grid-side power* P_{gr} *variation with time. Legend:* P_{gr}*—blue trace,* P_{dc}*—red trace.*

4.2.2.1 DC link voltage

The first extracted characteristic is shown in **Figure 14**. It illustrates the variation of the DC link and grid-side powers, P_{dc} and P_{gr}, against V_{dc}. The latter reflects the programmed controlled trajectory, which is also shown in **Figure 14**. The close match between the two traces verifies that the wind system was tightly controlled by the control system. The operating points are seen to spread over the entire characteristic, reflecting the higher wind speeds, as shown in **Figure 13**.

The variation of the DC link current I_{dc} with respect to the DC link voltage V_{dc} is shown in **Figure 15**. The current initially increased with the voltage as the power increased. This continues until a peak current is reached, at approximately 300 V. Comparison with **Figure 14** identifies that the voltage marks the end of region II. As the operation enters the constant power region, the trend of I_{dc} changes and starts to drop with increasing voltage. The slope at which the current decreased was less than the initial rate of increase.

The operation of the brake circuit is illustrated in **Figure 16**. Owing to the fast operation of the hysteresis controller, instantaneous values subsampled at 1 kHz were used. The higher frequency accounts for the ripple visible in the plot.

As the DC link voltage V_{dc} exceeds the upper threshold (approximately 380 V) following an operation in the constant power region, the brake circuit is activated. With the introduction of the resistive load, the PMG power approaches 4.5 kW, that is,

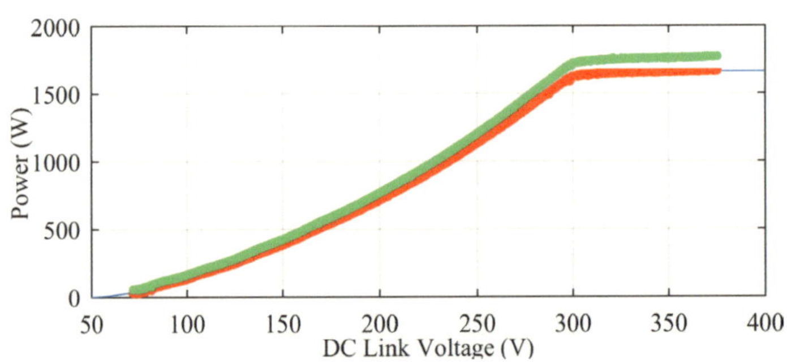

Figure 14.
Variation of the DC link power P_{dc} *and grid power* P_{gr} *with the DC link voltage* V_{dc}*. Legend: Control characteristic—blue trace,* P_{gr}*—red trace,* P_{dc}*—green trace.*

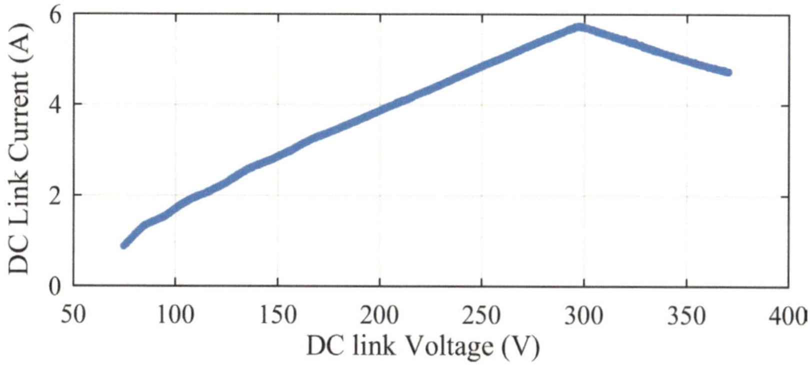

Figure 15.
Variation of the DC link current I_{dc} *with DC link voltage* V_{dc}.

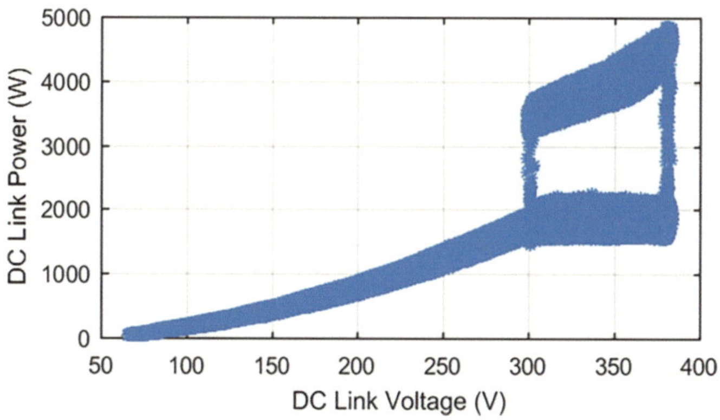

Figure 16.
Overall DC link power P_{dc} *variation with DC link voltage* V_{dc}.

250% of the rated power. The additional generator electromagnetic torque effectively opposes the wind turbine rotor torque, thereby reducing the rotational speed n_g. The reduction in n_g translates to a drop in V_{dc} and a corresponding decrease in the brake circuit power. The operation of the hysteresis controller is clearly shown in **Figure 16**, where the DC link voltage drops to the lower threshold for the brake circuit to be deactivated. During the brake operation, the power flow to the inverter stage continues as in the constant power region.

Overspeed protection is facilitated by the marked overload capability of the PMG. The increased current flow in the generator has a marked effect on the heat dissipation in the windings; thus, the braking operation can only be sustained for short periods of time. Depending on the relative size of the PMG with respect to the turbine rotor and the wind climate in which the system operates, an additional means of overspeed protection, such as active furling, independent of the generator, may be necessary to protect against persistent higher wind speeds.

4.2.2.2 Rotational speed

The second set of characteristics was extracted based on the generator rotational speed. This is an important parameter for system operation. However, owing to the use of the rectifier stage, the rotational speed can only be controlled indirectly.

It has been previously observed that a relationship between the rotational speed and DC link voltage is commonly assumed. For these subsequent plots, the generator rotational speed was extracted from the frequency of the measured AC stage voltage.

The variation in the DC link power P_{dc} with the rotational speed n_g is shown in **Figure 17**. The characteristic is similar to that against V_{dc}, as shown in **Figure 14**. The figure identifies the corresponding speed threshold for entering the constant power region. Overall, it can be observed that the rotational speed defines a coarser relationship with P_{dc}. Therefore, it is interesting to examine the relationship between the DC link voltage and the rotational speed owing to its central role in the use of indirect control.

The extracted V_{dc} vs. n_g characteristic is shown in **Figure 18**. It can be observed that the DC link voltage closely follows the rotational speed in region II. However, the speed varies within a band for the same DC link voltage, where the band extends at higher voltage levels. The variation band indicates that other uncontrolled variables play a role in the translation from the DC link voltage to the rotational speed. As the operation entered the constant power region, the relationship exhibited a steeper slope. This indicates a change in the rectifier operation.

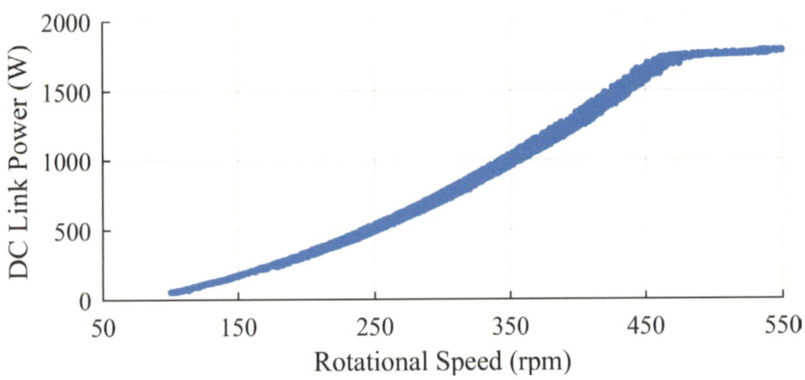

Figure 17.
Variation of the DC link power P_{dc} *with the generator rotational speed* n_g.

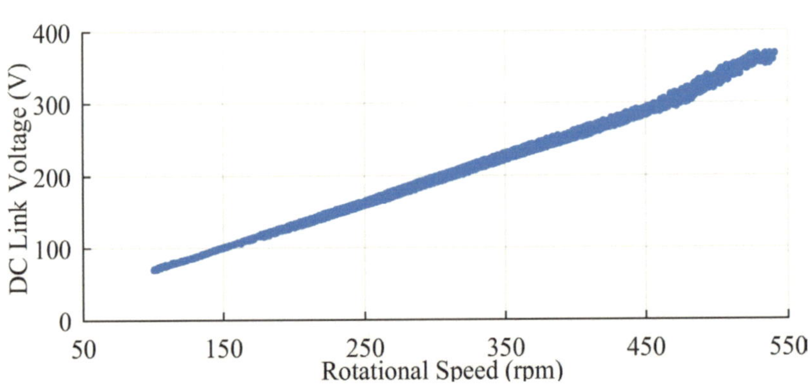

Figure 18.
Variation of the DC link voltage V_{dc} *with the generator rotational speed* n_g.

4.3 Temperature effects

Multiple runs were performed on each test day to examine the system behaviour in practical applications over extended periods. A gradual change in the system characteristics was observed, which was attributed to the increase in the generator temperature. This effect is elaborated in this section.

The progression of the DC link power P_{dc} against the rotational speed n_g characteristic for test day 2 is shown in **Figure 19**. The respective characteristics extracted from successive runs throughout the day are superimposed. Focusing on region II, it can be noted that the power per rotational speed initially decreases to eventually maintain a lower value. The effect is more evident as the rotational speed increases. This effect was not observed in the constant power region because the amount of power was directly controlled in this region.

To consolidate the observation, similar characteristics from the other two test days are plotted in **Figure 20**. As observed previously, the experienced wind speed was lower on both days. Considering **Figure 20(a)**, the lower wind speeds on test day 1 can be immediately noticed as the characteristic only covers part of the control trajectory. The successive tests were also run over a shorter period than that shown in **Figure 19**. Despite these conditions, a decrease in the DC link power P_{dc} is also observed. The effect is also visible in the characteristics from test day 3, as shown in **Figure 20(b)**. The observed effect reflects the conditions that are between those of the other test days, as previously noted.

The variation in P_{dc} is attributed to the increase in PMG temperature. The temperature variation depends on the effective cooling rate offered by the generator mounting arrangement. Given the indirect nature of the control, the reduction in power is explained as a variation in the relationship between V_{dc} and n_g with the generator temperature.

Similar to the case of the DC link power, successive DC link voltage V_{dc} vs. rotational speed n_g characteristics for test day 2 are superimposed in **Figure 21**. The overall shape of the characteristic is maintained over subsequent tests, but the specific

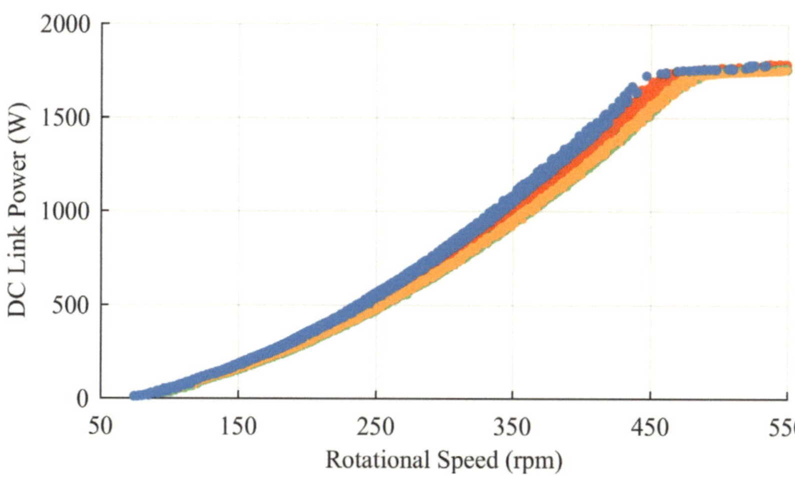

Figure 19.
Superimposed DC link power P_{dc} *vs. generator speed* n_g *characteristics showing the progression over test day 2. The run sequence follows the trace colours as follows: blue, red, green and orange.*

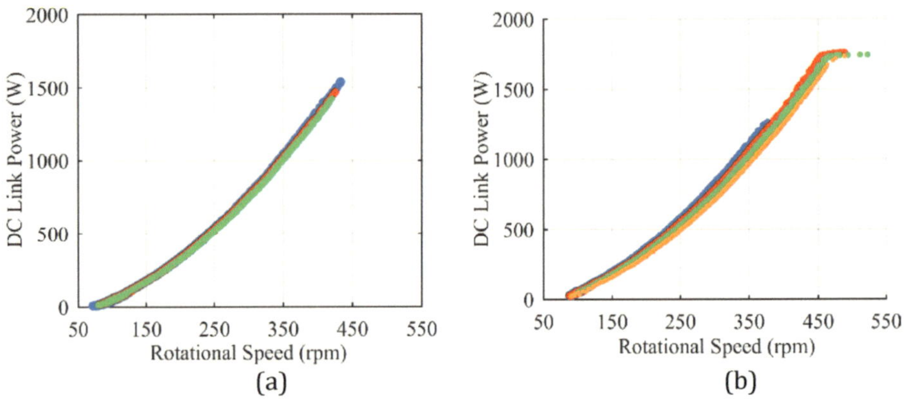

Figure 20.
Superimposed DC link power P_{dc} *vs. generator speed* n_g *characteristics showing the progression over (a) the test day 1 and (b) test day 3. The run sequence follows the trace colours as follows: blue, red, green and orange.*

V_{dc} to n_g translation shifts with the generator temperature. This, in turn, shifts the system operating point.

The inferred shift in the system operating point due to the generator temperature rise was verified by performing temperature tests on the wind-emulator rig. The rig was operated at constant rotational speeds, as in previous tests. As indicated earlier, the maximum P_{dc} was restricted in accordance with the capability of the rig. The system was left to operate until the generator frame temperature reached 60°C. Heat is generated through losses within the generator. For the generator used, the copper losses were significant. The observed P_{dc} variations at rotational speeds of 300 and 400 rpm are superimposed on the control trajectory, as shown in **Figure 22**. As the generator temperature increases, the operating points move towards the left, leading to a drop in P_{dc}. It can be observed that the control trajectory regulates the system operation, with the operation point variation as the generator temperature rises being dictated by the control trajectory itself. The operating points slide towards lower V_{dc} levels on the programmed control trajectory as the generator temperature increases.

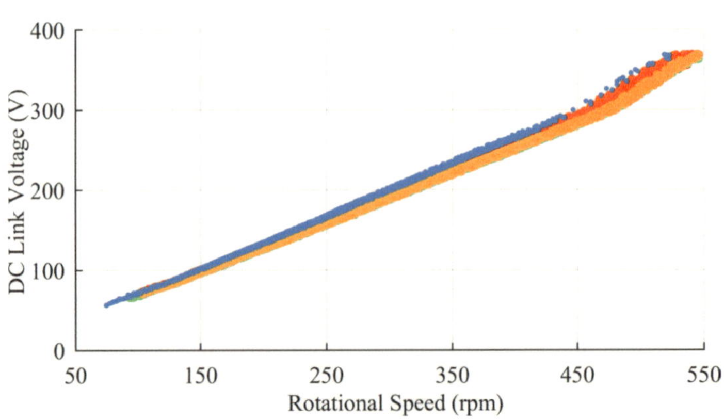

Figure 21.
Superimposed DC link voltage V_{dc} *vs. generator speed* n_g *characteristics showing the progression over test day 2. The run sequence can be identified through the trace colours, as follows: blue, red, green and orange.*

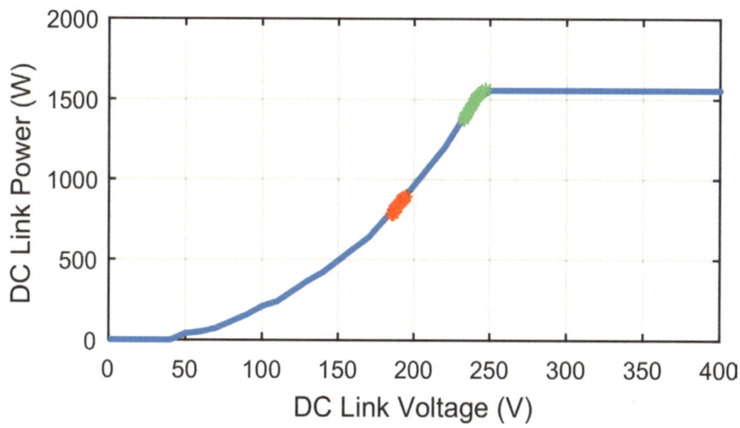

Figure 22.
Superposition of the observed P_{dc} *variations on the control trajectory for the 300 and 400 rpm controlled-temperature tests. Legend: 300 rpm—green trace, 400 rpm—green trace.*

5. Conclusion

This chapter presented the derivation and implementation of a control characteristic for a small-scale wind-energy conversion system with a nine-bladed rotor and a permanent magnet generator (PMG). The power conversion topology includes a diode rectifier, a DC-DC converter and a grid-connected inverter stage. The control characteristic is stored as a one-dimensional lookup table relating the grid-side power to the DC link voltage.

The derivation process covers operations in regions II and III of the wind turbine power curve. In region II, the control aims to extract maximum power while considering the lower-rated generator. The generator torque characteristic is used as a starting point, and suitable quasi steady-state operating points are identified at the intersection of the wind turbine aerodynamic power curves. These points are then translated to DC link variables using a simplified software model. In region III, the extracted power is limited to the generator-rated value by increasing the rotational speed above the nominal value. A brake circuit is used to protect against excessive rotational speeds.

Experimental tests were conducted on a wind-emulator rig under controlled laboratory conditions and on a prototype multi-bladed wind system in the field. The instantaneous waveforms exhibited significant complexity. Field tests over multiple days with varying wind-speed conditions demonstrated the system's operation across the entire control characteristic range. Selected characteristics against the DC link voltage and the rotational speed were examined.

Temperature effects were observed during extended operation, with the power per rotational speed decreasing as the generator temperature increased. This was attributed to a shift in the relationship between DC link voltage and rotational speed with rising generator temperature. The control trajectory regulates the system operation, with the operating point sliding towards lower voltage levels as temperature increases.

Nomenclature

ρ	air density (kg/m^3)
A	rotor cross-sectional area (m^2)

R	turbine rotor radius (m)
C_p	power coefficient
λ	tip speed ratio
β	blade pitch angle (°)
V_w	wind speed (m/s)
ω_{wt}	rotor rotational speed (rad/s)
n_{wt}	rotor rotational speed (rpm)
P_{gi}	generator input power (W)
P_{go}	generator output power (W)
P_{wt}	wind-turbine-extracted-power (W)
τ_g	generator torque (Nm)
n_g	generator rotational speed (rpm)
J	moment of inertia (kgm^2)
G	gearbox ratio
P_{ac}	AC-side power (W)
P_{dc}	DC link power (W)
P_{gr}	grid-side power (W)
V_{ac}	AC-side line voltage (V)
V_{dc}	DC link voltage (V)
V_{gr}	grid-side voltage (V)
I_{ac}	AC-side phase current (A)
I_{dc}	DC link current (A)
I_{gr}	grid-side current (A)
V_{sd}	D-axis stator voltage (V)
V_{sq}	Q-axis stator voltage (V)
i_{sd}	D-axis stator current (A)
i_{sq}	Q-axis stator current (A)
L_s	stator inductance (H)
R_s	stator resistance (Ω)
ω_e	generator electrical frequency (rad/s)
ψ_{pm}	permanent magnet flux (Wb)
ω_m	generator mechanical frequency (rad/s)
p	pole pairs

Author details

Cedric Caruana* and Claire Grima
Faculty of Engineering, Department of Electrical Engineering, University of Malta, Malta

*Address all correspondence to: cedric.caruana@um.edu.mt

References

[1] Global Wind Report 2025 [Internet]. 2025. Available from: https://www.gwec.net/reports/globalwindreport [Accessed: 1 August 2025]

[2] WWEA Annual Report 2024 [Internet]. 2024. Available from: https://wwindea.org/AnnualReport2024 [Accessed: 1 August 2025]

[3] Distributed Wind Market Report: 2021 Edition [Internet]. 2021. Available from: https://www.energy.gov/eere/wind/articles/distributed-wind-market-report-2021-edition-released [Accessed: 1 August 2025]

[4] Bianchini A et al. Current status and grand challenges for small wind turbine technology. Wind Energy. 2022;7:2003-2007. DOI: 10.5194/wes-7-2003-2022

[5] Small Wind Turbine Market [Internet]. 2025. Available from: https://market.us/report/global-small-wind-turbine-market [Accessed: 1 August 2025]

[6] Small Wind Market Forecasts to 2030 [Internet]. 2024. Available from: https://www.giiresearch.com/report/smrc1530668-small-wind-market-forecasts-global-analysis-by.html [Accessed: 1 August 2025]

[7] Barzegar-Kalashani M et al. Small-scale wind turbine control in high-speed wind conditions: A review. Sustainable Energy Technologies and Assessments. 2023;**60**(103577):1-17. DOI: 10.1016/j.seta.2023.103577

[8] Chen J, Chen J, Gong C. New overall power control strategy for variable-speed fixed-pitch wind turbines within the whole wind velocity range. IEEE Transactions on Industrial Electronics. 2013;**60**(7):2652-2660. DOI: 10.1109/TIE.2012.2196901

[9] Dalala ZM, Zahid ZU, Lai J-S. New overall control strategy for small-scale WECS in MPPT and stall regions with mode transfer control. IEEE Transactions on Energy Conversion. 2013;**28**(4):1082-1092. DOI: 10.1109/TEC.2013.2287212

[10] Lumbreras C et al. Control of a small wind turbine in the high wind speed region. IEEE Transactions on Power Electronics. 2016;**31**(10):6980-6991. DOI: 10.1109/TPEL.2015.2508674

[11] Li W, Abbey C, Joos G. Control and performance of wind turbine generators based on permanent magnet synchronous machines feeding a diode rectifier. In: IEEE Power Electronics Specialist Conference; 18-22 June 2006; Korea (South). New York: IEEE; 2006. pp. 1-6

[12] Knight AM, Peters GE. Simple wind energy controller for an expanded range. IEEE Transactions on Energy Conversion. 2005;**20**(2):459-466. DOI: 10.1109/TEC.2005.847995

[13] Caruana C, Grima C, Mifsud M. Performance aspects of the electrical conversion stages of a small multibladed wind energy system. In: IEEE IAS Global Conf. on Renewable Energy and Hydrogen Technologies (GlobConHT); 11-12 March 2023; Male. New York: IEEE. pp. 1-6

[14] Kao J-H. Developing the process in matching a win turbine system to attain optimal performance. Advances in Mechanical Engineering. 2016;**8**(11):1-9. DOI: 10.1177/1687814016674697

[15] Eriksson S. Direct driven generators for vertical axis wind turbines [thesis]. Uppsala: Uppsala University; 2008

[16] Tan K, Islam S. Optimum control strategies in energy conversion of PMSG wind turbine system without mechanical sensors. IEEE Transactions on Energy Conversion. 2004;**19**(2):392-399. DOI: 10.1109/TEC.2004.827038

[17] Raza Kazmi SM et al. Review and critical analysis of the research papers published till date on maximum power point tracking in wind energy conversion system. In: Proceedings of the 2010 IEEE Energy Conversion Congress and Exposition; 12-16 September 2010; Georgia. New York: IEEE; 2010. pp. 4075-4082

[18] Brando G, Coiro DP, Dannier A. An efficient power control strategy for small fixed-pitch wind turbine to extend the operating range to high wind speed region. In: Proceedings of the Int. Symposium on Power Electronics, Electrical Drives, Automation and Motion (SPEEDAM2012); 20-22 June 2012; Italy. New York: IEEE; 2012. pp. 541-546

[19] Shafiei A et al. A control approach for a small-scale PMSG-based WECS in the whole speed range. IEEE Transactions on Power Electronics. 2017;**32**(12):9117-9130. DOI: 10.1109/TPEL.2017.2655940

[20] Ashraf A, Ran L, Bumby JR. New constant electrical power soft-stalling control for small-scale VAWTs. IEEE Transactions on Energy Conversion. 2010;**25**(4):1152-1161. DOI: 10.1109/TEC.2010.204373

[21] Zhu Y et al. A novel maximum power point tracking control for permanent magnet direct drive wind energy conversion systems. Energies. 2012;**5**:1398-1412. DOI: 10.3390/en5051398

[22] Yamamura N, Ishia M, Hori T. A simple wind power generating system with permanent magnet type synchronous generator. In: Proceedings of the IEEE 1999 Int. Conference on Power Electronics and Drive Systems (PEDS '99); 27-29 July 1999; Hong Kong. New York: IEEE; 1999. pp. 849-854

[23] Kumari S, Kushwaha V, Gupta TN. A maximum power point tracking for a PMSG based variable speed wind energy conversion system. In: Proceedings of the 2018 Int. Conf. on Power, Energy, Environment and Intelligent Control (PEEIC); 13-14 April 2018; India. New York: IEEE; 2018. pp. 789-794

[24] De Broe AM, Drouilhet S, Gevorgian V. A peak power tracker for small wind turbines in battery charging applications. IEEE Transactions on Energy Conversion. 1999;**14**(4):1630-1635. DOI: 10.1109/60.815116

[25] Stannard N, Bumby JR. Performance aspects of mains connected small-scale wind turbines. IET Generation Transmission and Distribution. 2017;**1**(2):348-356. DOI: 10.1049/iet-gtd:20060167

[26] Borunda M et al. Intelligent control of an experimental small-scale wind turbine. Energies. 2024;**17**(15656):1-31. DOI: 10.3390/en17225656

[27] Salih BM, Hamoodi SA, Hamoodi AN. Enhancing the behaviour of small scale wind turbine based on fuzzy logic system. Journal European des Systemes Automatises. 2024;**57**(1):147-153. DOI: 10.18280/jesa.570115

[28] Zhou Z et al. Power control of a nonpitchable PMSG-based marine current turbine at overrated current speed with flux-weakening strategy. IEEE Journal of Oceanic Engineering. 2015;**40**(3):536-545. DOI: 10.1109/JOE.2014.2356936

[29] Boldea I. Permanent magnet synchronous generator systems. In: Boldea I, editor. Variable Speed Generators. 2nd ed. Boca Raton: CRC Press; 2015. pp. 10-1-10-86. DOI: 10.1201/b19293

[30] Sant T et al. Development and performance testing of a small, multibladed wind turbine. Wind Engineering. 2019;**44**(1):3-20. DOI: 10.1177/0309524X19849845

[31] Gerlando AD et al. Comprehensive steady-state analytical model of a three-phase diode rectifier connected to a constant DC voltage source. IET Power Electronics. 2013;**6**(9):1927-1938. DOI: 10.1049/iet-pel.2013.0026

[32] Caruana C, Grima C. Study on the power generation variation of small-scale wind systems due to PMG temperature rise. In: 2023 IEEE Int. Conference on Energy Technologies for Future Grids (ETFG); 03-06 December 2023; Australia. New York: IEEE; 2024. pp. 1-6